W9-BKB-706

Three separate Voyager 1 images taken through ultraviolet, green and violet filters were used to construct this color enhanced composite of Saturn. (Courtesy Jet Propulsion Laboratory/ NASA.)

ATMOSPHERES

ATMOSPHERES

A View of the Gaseous Envelopes Surrounding Members of Our Solar System

James P. Barbato
Elizabeth A. Ayer

Pergamon Press
New York • Oxford • Toronto • Sydney • Paris • Frankfurt

Pergamon Press Offices:

U.S.A.	Pergamon Press Inc., Maxwell House, Fairview Park, Elmsford, New York 10523, U.S.A.
U.K.	Pergamon Press Ltd., Headington Hill Hall, Oxford OX3 0BW, England
CANADA	Pergamon Press Canada Ltd., Suite 104, 150 Consumers Road, Willowdale, Ontario M2J 1P9, Canada
AUSTRALIA	Pergamon Press (Aust.) Pty. Ltd., P.O. Box 544, Potts Point, NSW 2011, Australia
FRANCE	Pergamon Press SARL, 24 rue des Ecoles, 75240 Paris, Cedex 05, France
FEDERAL REPUBLIC OF GERMANY	Pergamon Press GmbH, Hammerweg 6, Postfach 1305, 6242 Kronberg/Taunus, Federal Republic of Germany

Library of Congress Cataloging in Publication Data

Barbato, James P. (James Paul), 1943-

Atmospheres, a view of the gaseous envelopes surrounding members of our solar system

Includes index.
1. Planets--Atmospheres. I. Ayer, Elizabeth A. (Elizabeth Anne). 1954- II. Title.
QB603.A85B37 1981 551.5'0999'2 81-1857
ISBN 0-08-025583-3 AACR2
ISBN 0-08-025582-5 (pbk.)

Printed in the United States of America

DEDICATION

To my mother and father, my wife, Ann and my sons, Brian, Steven, and Mark and

in memoriam

my good friend, Ken Cardner, whose encouragement in the early stages of the development of this manuscript meant so much to me.

J.P.B.

To my parents, to my sisters, Nancy, Cindy, Gretchen, and to my friends -- for their love, encouragement, and understanding.

E.A.A.

TABLE OF CONTENTS

COLOR PLATES

LIST OF ILLUSTRATIONS

CHAPTER I

CHAPTER IV

CHAPTER V

LIST OF SYMBOLS AND ABBREVIATIONS

A.U.	astronomical unit, the distance from the earth to the sun (150,000,000 km)
bar	unit of atmospheric pressure measurement equal to a force of 1 million dynes per square cm per minute
cal	calories
cm^{-3}	cubic centimeters
$g\ cal\ cm^{-2}\ min^{-1}$	gram calorie per square centimeter per minute
$g\ cm^{-3}$	grams per cubic centimeter
Hz	hertz (formerly cycles per second)
^{o}K	degrees Kelvin
$^{o}K\ km^{-1}$	degrees Kelvin per kilometer
$km\ hr^{-1}$	kilometers per hour
$km\ sec^{-1}$	kilometers per second
ly	langly: 1 ly = 1 $g\ cal\ cm^{-2}\ min^{-1}$
m^{-3}	cubic meters
mb	millibar (1/1,000 of a bar)
$mg\ m^{-3}$	microgram per cubic meter
$m\ sec^{-1}$	meters per second
P.A.L.	present atmospheric level
ppm	parts per million
pr μm	precipitable micrometers (of water)
R_E	earth radii, R_J - Jupiter radii, etc.
$volts\ cm^{-1}$	volts per centimeter
$W\ m^{-2}$	watts per square meter
$>$	greater than
$<$	less than
$\simeq$	approximately, approximately equal to
μm	micrometers, 1 μm = 0.0001 cm

PREFACE

This book is based on the new information obtained by the Voyager 1 and 2 as well as the Viking, Pioneer, and Mariner space probes. It is designed for college students, who may have already taken an introductory meteorology or astronomy course, and amateur meteorologists and astronomers, world-wide.

The book is written so that most of the scientific terminology is introduced and explained in chapter one. Chapter one summarizes the processes that occur in the earth's atmosphere. A mastery of the first chapter should allow the reader to proceed easily through the remainder of the book. The following chapters are sequentially ordered based on increasing heliocentric distances beginning with Mercury.

Each chapter attempts to discuss certain key topics such as the composition of the atmosphere, the vertical temperature/pressure structure, the energy balance, the magnetosphere, and meteorological phenomena (general circulation/winds, clouds, etc.). One or more of these topics may not appear in several chapters, not because of omissions by the authors but rather because the data necessary to develop these areas are non-existent.

Atmospheres is written in a descriptive style, and is designed as a primer on planetary atmospheres. Selected journal references are provided at the end of each chapter.

There were occasions during the compilation of this book when we wondered whether we were writing a scientific text or a scientific "fairy tale." So often, the opening sentences of a paragraph dealt with hard scientific evidence, only to be followed by a series of sentences with verbs such as "imply," "suggest," "may be," or "appears that." These words truly reflect the "state of the art" in the study of planetary atmospheres where one piece of new data raises several direct or peripheral questions, the answers to which are inference, conjecture, or speculation at the present time.

Sterling, Mass.
Bartlesville, Ok.
January, 1981

James P. Barbato
Elizabeth A. Ayer

ACKNOWLEDGEMENTS

The authors wish to acknowledge the scientists associated with the Jet Propulsion Laboratory of the California Institute of Technology and NASA. Their research and data have made the writing of this book possible.

Alan Bourgault contributed his cartographic talents. Collette Flynn contributed her typing skills in doing the rough copy. Gloria Sieniuc contributed her talents in editing and typing the final copy. Lastly, we are indebted to Clarence E. Brideau and Louis Manring for their tireless efforts in the preparation of the index and proofreading the manuscript.

J.P.B.

E.A.A.

CHAPTER I

THE EARTH

Descriptive Statistics:

Diameter	
Equatorial	12,757 km
Polar	12,714 km
Period of Rotation	24 Hours
Distance from the Sun	1.0 A.U. (150,000,000 km)
Surface Gravity	1.0 (the standard against which the other planets are compared)
Mean Density	5.5 g cm^{-3}
Mass	1.0 (the standard against which the other planets are compared)

Introduction

There is general agreement that the earth's primordial atmosphere formed $\simeq 4.5 \times 10^9$ years ago. The theories for its origin usually postulate either condensation from hot gases or from the aggregation of cold cosmic dust. Whatever theory one chooses, most agree that the earth was hot and molten after its formation due to the pressure and heat of gravitational contraction. Any atmosphere the earth may have had at this point quickly escaped into space, in the same way a boiling liquid evaporates. High atmospheric temperatures are necessary to explain the small amounts of light, volatile elements such as helium, argon, neon, and free molecular hydrogen on earth. The high temperatures are also necessary to explain the relatively small amounts of the heavy noble gases, xenon and krypton as compared to the amounts of these gases found in the sun and the rest of the universe. High temperatures would provide the kinetic energy needed for their (xenon and krypton) escape. It is also possible that most of the xenon and krypton was lost before the actual aggregation of the earth occurred. The small particles of cosmic dust with their low gravitational attraction probably could not hold much of these heavier, chemically unreactive gases. The temperature increase at the earth's formation occurred simultaneously with an increase in the mass of the earth and greater gravitational attraction. This larger gravitational field made it possible to retain volatile compounds of carbon, nitrogen, and oxygen. The reason our atmosphere does not contain certain gases, especially the noble gases, in the same ratio as the rest of the universe and our sun, can be explained by one of the preceding ideas or some combination of them.

The early atmosphere, one that the earth could retain, was highly reduced, i.e. materials had lost electrons. The atmosphere probably consisted of methane (CH_4), with lesser amounts of ammonia (NH_3), molecular hydrogen (H_2), and water vapor prior to the migration of the metallic iron into the core. Free oxygen was absent at this stage. The best estimates place the formation of the earth's core at $\simeq 3.8 \times 10^9$ years ago. The chemical stability of an early methane-ammonia atmosphere is an unresolved question. It may have lasted from only 10^5-10^7 years to as long as 10^8 years.

The formation of the earth's metallic core had the effect of increasing the average degree of oxidation of materials outside the core because the core formation removed free metal from the material which solidified to form the mantle and the crust. After the core formation, the outgassed volatiles probably became less reduced. Methane (CH_4), ammonia (NH_3), and hydrogen (H_2) were replaced by an atmosphere of carbon dioxide (CO_2), nitrogen (N_2), and water vapor ($H_2O_{(v)}$), respectively. The final composition of this early atmosphere 3.0-3.8 x 10^9 years ago consisted of a reducing mixture of water vapor ($H_2O_{(v)}$), molecular hydrogen (H_2), nitrogen (N_2), carbon dioxide (CO_2), methane (CH_4), ammonia (NH_3), and some trace components. The exact composition of this mixture is uncertain, however, free oxygen (O_2) was not present and no ozone (O_3) was present at higher atmospheric levels to shield the earth from high-energy ultraviolet radiation.

Most of the ideas concerning the primordial atmosphere are speculative and are based on our limited knowledge of the composition of the universe and the manner in which chemical species behave. As we approach the more recent history of the earth's atmosphere, any hypothesis presented must agree with the tangible evidence. This evidence is in the form of rocks in the earth's lithosphere. Certain types of rocks and associated mineral assemblages place limits on the composition and conditions of the atmosphere beneath which they were deposited. This is true even today. Calcite ($CaCO_3$) is not deposited in basic or acidic environments, which implies that limestone ($CaCO_3$), and dolomite ($CaMg(CO_3)_2$), common sedimentary materials, are deposited in aqueous environments with near neutral pH values. Since surface waters contain dissolved gases from the atmosphere, a highly reducing or oxidizing atmosphere would not favor the precipitation of calcium carbonate.

Rocks older than 1.8-2.0 x 10^9 years consist of abundant, bedded chemical silicates which make up the chert component of the banded iron formations found with little or no limestone and dolomite. This suggests that there could have been only small quantities of ammonia in the atmosphere for ammonia would raise the pH of the hydrosphere and favor an abundance of carbonate rocks. Bedded cherts would be a rarity. The deposition of carbon of a nonvital origin would have resulted had methane been present in large quantities. Although some nonvital carbon is found in sedimentary rocks, it is not a conspicuous component of these oldest sediments. This evidence suggests that by 2.5 x 10^9 years, atmospheric levels of methane and ammonia were low. The sedimentary rocks lend support to the hypothesis that the atmosphere consisted of gases occluded in igneous rocks or recognized as juvenile gases emitted primarily during volcanic eruptions. These gases are water vapor, carbon dioxide, carbon monoxide, nitrogen, sulfur dioxide, hydrogen chloride, and a few other trace gases.

The composition of the atmosphere began to change with the appearance of oxygen-releasing photosynthesizers. These early photosynthesizers were faced with the problem of disposing of the free oxygen (O_2) they had produced in a way as not to oxidize (burn) themselves. The abundance of hematitic banded iron formations among sediments deposited between 1.8-2.0 x 10^9 years suggest that ferrous iron acted as an oxygen acceptor in the physical environment. The banded iron formations represent extensive deposition cycles in large open bodies of water. Alternating layers of iron-rich and iron-poor silica were deposited by these cycles. Nothing matching their regional extent is found in younger rocks. In younger rocks, iron deposition appears to be mainly chemical replacement, geochemically a very different source. The geochemical problem of the banded iron formations is how to explain the transport of the iron in solution under oxidizing conditions. Cloud (1968) has proposed a balanced relationship between primitive green plant photosynthesizers and the banded iron formation that may resolve the problem. The iron could be transported in solution in the ferrous state and precipitated as ferric or ferro-ferric iron upon combination with the oxygen from photosynthesis. The cyclic banding could result from a delicately fluctuating balance between the abundance of green plant photosynthesizers and the supply of ferrous iron. The oxygen released by the plants became bound and precipitated as a chemical sediment from the hydrosphere and, except for very small quantities, did not leak into the atmosphere. These small quantities were rapidly scavenged by the reduced substances in the atmosphere and at the surface of the lithosphere. Free oxygen probably did not exceed 10^{-3} P.A.L. (present atmospheric level). Its source would have been the photodissociation of water vapor and carbon dioxide.

This delicate balance between plants and available iron would have terminated as more advanced photosynthesizers, equipped to cope with oxygen and peroxide, evolved. These advanced photosynthesizers became widespread and eliminated free ferrous iron from the hydrosphere. The evidence points to one last episode of banded iron formation 1.8-2.0 x 10^9 years B.P. (before the present). This last episode of banded iron formation coincides with a marked increase in blue-green algae.

The oxygen produced by the algae probably saturated the hydrosphere and began to invade the atmosphere. Ultraviolet energy dissociated some of the molecular oxygen (O_2) into atomic oxygen (O^*) and ozone (O_3). Atomic oxygen and ozone are highly reactive, even at low concentrations. Their presence in the atmosphere initiated a period of rapid oxidation of surface materials. Available records show that the oldest thick and extensive red beds, detrital continental, or marginal marine sediments in which the individual grains are coated with ferric oxides, are about 1.8-2.0 x 10^9 years old. They are a little younger than, or overlap slightly with, the most recent banded iron formation. This lithospheric event may mark one of the most significant periods in the history of the earth's atmospheric evolution, the time when free oxygen began to accumulate.

Ultraviolet penetration into the hydrosphere was probably 5-10 m during this period. It is unlikely that plant photosynthesizers inhabited the upper regions of the oceans until the ozone layer in the stratosphere was well established. The composition of the earth's atmosphere by 600 million years ago during the Precambrian probably closely resembled today's atmosphere.

The earth has a gaseous envelope very different from its two closest neighbors, Mars and Venus. Certain conditions on earth are not what would be expected for a planet interpolated between Mars and Venus. Lovelock and Margulis (1974) have proposed an interesting concept, the Gaia Hypothesis, to explain these differences. This hypothesis puts forth the idea that the biosphere actively maintains optimum conditions for life. The atmospheric composition, oxidation-reduction potential, and temperature are modulated by and for the biosphere. One anomaly that they use to support their hypothesis is the existence of gaseous nitrogen (N_2) in the atmosphere when the chemically stable form of the element nitrogen on earth is the nitrate ion in solution in the oceans. Another chemical anomaly is the presence of oxygen (O_2) and methane (CH_4), both of which are highly reactive. These gases, they argue, would not be found in the atmosphere if life was not present on our planet and regulating the composition of the atmosphere. Lovelock and Margulis feel that the composition of the atmosphere is far from any conceivable abiological steady state equilibrium and is more consistent with the idea that it is a mixture of gases for some specific purpose, i.e. the existence of life.

Lovelock and Margulis also suggest that if life were to cease, levels of oxygen and nitrogen would decline until they were trace gases in an atmosphere of water vapor, carbon dioxide, and the noble gases. Without life, earth would have an atmosphere whose composition was something of a compromise between the atmospheres of Venus and Mars. This assumption may be correct, but the proponents of the Gaia Hypothesis seem to want to link the biosphere with the rest of the planet as a "controlling entity." There is little doubt that a strong mutual interdependence and series of feedback mechanisms exist between the biosphere and the atmosphere. The entire Gaia controversy centers around the evolution of life. Is it not more reasonable to think of the conditions on earth as a fortuitous accident of chance? The planet's orbital characteristics and distance from the sun permit the existence of a temperature regime where water may exist in all three of its phases, i.e. solid, liquid, and gas. The high temperatures on Venus permit only water vapor to exist. The cold thermal regime of Mars only permits transformations from vapor to solid water (ice). Liquid water is generally considered an absolutely essential ingredient for life. The earth and its vast oceans were therefore "primed" for the evolution of life forms.

Recent geochemical evidence (Kerr, 1980) suggests that a reducing atmosphere of hydrogen, methane, and ammonia was probably short-lived on the earth. Volcanic outgassing contributed carbon dioxide, nitrogen, hydrogen, oxygen (though not in amounts currently present), and minor gases such as hydrogen sulfide, etc. Carbonate minerals are found in rocks $\approx 3.8 \times 10^9$ years old. The formation of rocks with carbonate minerals requires an atmosphere of carbon dioxide, not methane. This would suggest that the transition from a strongly reducing to an oxidizing atmosphere occurred at least prior to the formation of these oldest known rocks.

Early theorists attempted to prove that the primordial atmosphere consisted of ammonia and methane by experiments. Electric arcs were passed through an atmosphere rich in ammonia and methane. Organic compounds (amino acids), the building blocks of life, were produced in large quantities. This experiment was considered proof for a reducing atmosphere when life began. It is now known that amino acids may be generated by electric discharges passed through laboratory atmospheres of carbon monoxide, nitrogen, and hydrogen and/or water although not in large quantities. Current views, reflecting the geochemical data, now suggest the earth's redox potential (i.e. reducing-oxidizing) was intermediate when life began on earth.

The Gaia Hypothesis, whether one accepts it or not, has been extremely valuable. It has stimulated renewed interest in primordial atmospheres and the origins of life. It has elevated the role of the biosphere and brought its interactions with the atmosphere to the attention of atmospheric scientists. The question of whether the biosphere "consciously" modulates the atmosphere for the preservation of life on earth remains an unanswered question. As we turn our attention to the study of the other planetary atmospheres, let us pay particular attention to our closest neighbors, Venus and Mars. Does life on earth account for the differences between the earth's atmosphere and that of Venus and Mars? Or, did random chance provide the fortuitous combination of thermal regime and chemical composition to permit the evolution of life only on the earth?

Composition of the Atmosphere

Ninety-nine percent of the earth's atmosphere consists of nitrogen (N_2) and oxygen (O_2). The remaining permanent constituents consist of the five noble gases, i.e. argon, neon, helium, krypton, and xenon as well as carbon dioxide (CO_2), hydrogen (H_2), methane (CH_4), nitrous oxide (N_2O), and radon (R_n) (Table 1.1). The variable

TABLE 1.1

The permanent constituents of the lower atmosphere (after Miller and Thompson, 3rd ed., courtesy of Charles E. Merrill Publishing Company, 1979, Page 3)

Constituents	Volume (%)
Major	
Nitrogen (N_2)	78.1
Oxygen (O_2)	20.9
Minor	
*Argon (Ar)	0.9
Carbon dioxide (CO_2)	0.033
*Neon (Ne)	18.0×10^{-4}
*Helium (He)	5.0×10^{-4}
*Krypton (Kr)	1.0×10^{-4}
*Xenon (Xe)	0.09×10^{-4}
Hydrogen (H_2)	$.5 \times 10^{-4}$
Methane (CH_4)	1.5×10^{-4}
Nitrous oxide (N_2O)	0.5×10^{-4}
Radon (Rn)	6.0×10^{-18}

*Delineates noble gases

constituents of the atmosphere are listed in Table 1.2. Water, the combination of hydrogen and oxygen, exists in its three physical states (phases) in the earth's atmosphere: liquid, solid, and gaseous. This is a result of the unique range of temperatures that exist on earth because of its distance from the sun. Many atmospheric constituents, both permanent and variable, i.e. water, carbon dioxide, ozone, and particulate matter, exert powerful influences on the behavior of the earth's atmosphere. The entire energy budget of our planet is modulated by the presence of certain gases, which, at first glance, appear insignificant (Table 1.2).

TABLE 1.2

Major variable constituents of the lower atmosphere. (After Miller and Thompson, Meteorology, 3rd ed., courtesy of Charles E. Merrill Publishing Company, 1979, page 3.)

Constituents	Volume (%)
Water vapor (H_2O)	<4
Carbon dioxide (CO_2)	0.033
Ozone (O_3)	$<0.07 \times 10^{-4}$
Particulates (dust, salt crystals)	$<10^{-5}$

The natural chemistry of our atmosphere is generally well understood by atmospheric scientists. The vast majority of the chemical reactions that occur in the atmosphere's upper layers are photochemical. Sunlight supplies the energy to dissociate molecules into atoms leaving them to recombine in various ways, depending on their chemical reactivity. The results of the recombinations of atoms influence the very nature and short term behavior of the earth's atmosphere, the characteristics of each of its layers, and the initiation of long period climatic cycles.

The atmosphere is divided into two major regions, the homosphere, or zone below 90 km, and the heterosphere, the zone between 90 km and the edge of space. The most significant atmospheric reactions within the homosphere involve the creation of the ozone layer and respectively, the ozonosphere. The relative concentration of ozone to air molecules (shown at the bottom left of Fig. 1.1), reaches a maximum at 30 km and decreases to a negligible concentration at altitudes between 55-65 km.

The Structure of the Earth's Atmosphere

The temperature/pressure relationships of the earth's atmosphere demonstrate a readily understandable pattern. Mean surface pressure at sea level is 1013.2 mb. It decreases vertically at a constant rate of one-thirtieth (1/30) of its value for every 275 m in altitude (Fig. 1.2). The vertical temperature profile does not depict a uniform temperature decrease with height. The temperature curve varies markedly in conjunction with the four major layers of the earth's atmosphere: the troposphere, stratosphere, mesosphere, and thermosphere. In general, the temperature profile is responsive

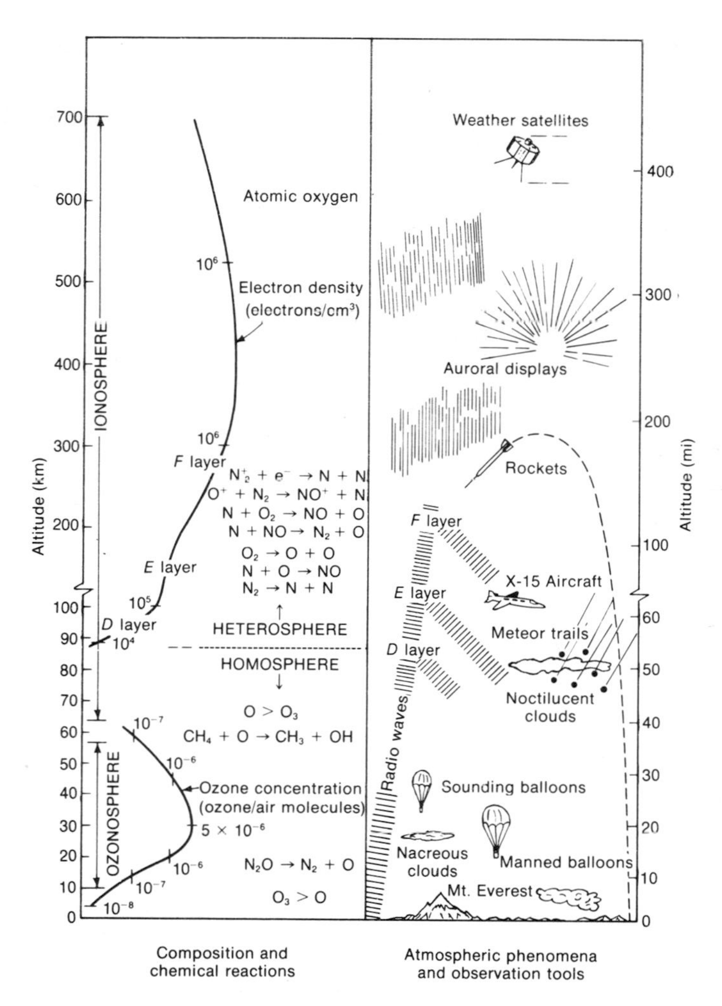

Fig. 1.1. Vertical distribution of atmospheric properties. (Courtesy Miller & Thompson, Meteorology, 3rd Ed., Charles E. Merrill Publ. Co., 1979, p. 5.)

to the way the earth is warmed by the sun. Energy penetrates the atmosphere and warms the earth's surface. The surface, in turn, radiates long wave, infrared heat energy. The atmosphere's gases absorb this heat energy which warms the atmosphere. Thus, the earth's atmosphere is heated from the bottom upward. Logically, one would expect air temperature to decrease with increasing altitude much the same as it does when one ascends a high mountain. The rate of temperature change with height is called the lapse rate. It is clearly evident from the temperature profile in Fig. 1.2 that temperatures increase and decrease with height above sea level. The temperature trend is easily determined by the direction the curve takes. The temperature increases as the curve swings to the right of the diagram and decreases as the curve swings to the left. Two

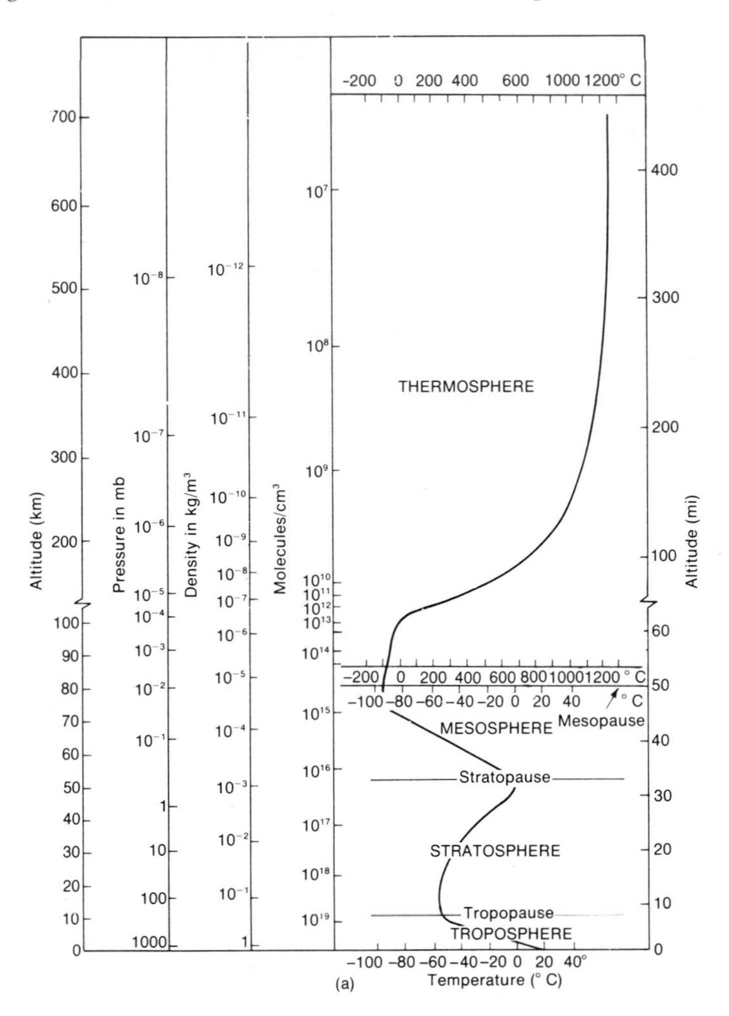

Fig. 1.2. Vertical temperature profile and layers of the atmosphere. (Courtesy Miller & Thompson, Meteorology, 3rd Ed., Charles E. Merrill Publ. Co., 1979, p. 4.)

layers, the troposphere and mesosphere, may be considered "normal" as they represent the expected effect, i.e. a temperature decrease with height.

The rate of temperature decrease with height in the troposphere, assuming that the earth's atmosphere is not in motion, is the environmental lapse rate. This value would be $\approx 6.5^{\circ}K\ km^{-1}$. While this value is interesting for comparative purposes, in reality, the earth's atmosphere is always in motion. The dry adiabatic lapse rate ($10^{\circ}K\ km^{-1}$) and the wet adiabatic lapse rate ($6^{\circ}K\ km^{-1}$) are two other lapse rate values which are more realistic for use in studies of temperature changes related to atmospheric motions.

A parcel of air that is forced to rise is moving from a region of greater atmospheric pressure into a region of lesser pressure. The result is an expansion of the gas, i.e. a volume change. The gas molecules are now spaced farther apart. This results in a measurable temperature decrease. The adiabatic lapse rate defines the rate of temperature change. Adiabatic processes therefore involve volume and temperature changes as air parcels move vertically up or down into regions of lesser or greater atmospheric pressure.

Air, which is forced to rise, will cool at the dry adiabatic rate until it reaches its dew point temperature, i.e. that temperature at which the air becomes saturated with water vapor. If the air continues to rise and cool, the water vapor will condense out to form clouds, releasing large quantities of latent heat to the atmosphere in the process. The heat released is not sufficient to terminate the cooling of the rising air but it does slow down the rate of cooling to $6^{\circ}K\ km^{-1}$. This is known as the wet adiabatic lapse rate and applies to all situations where the condensation of water vapor occurs in rising columns of air. A discussion of latent heat transfer follows in the section discussing the earth's energy balance.

The earth's atmospheric temperature will generally decline with vertical ascent at a rate somewhere between 6°-$10^{\circ}K\ km^{-1}$ depending on whether condensation is occurring. Since most of the earth's weather occurs in the troposphere, adiabatic changes represent the initiation of atmospheric changes. Other planetary atmospheres may have similar adiabatic lapse rate patterns. The measurement and comparison of those rates with the lapse rates occurring in the earth's atmosphere allow scientific comparisions to be made.

Lapse rate data provide the means to analyze the vertical temperature characteristics of an atmosphere, particularly in the troposphere. The thermal characteristics of the air will determine its potential for motion and therefore define states of atmospheric stability, instability, or conditional instability. Stability may be defined as existing when the lapse rate in an air mass or parcel is less than the adiabatic lapse rate. These conditions generally prevail on cloudless days with bright sunshine, usually under conditions where the air is gently subsiding (sinking). Instability exists when the lapse rate of the parcel of air is greater than the adiabatic lapse rate. Absolute instability is not common but when it does exist, the lapse rate is less then the wet adiabatic lapse rate. Rapid changes occur when absolute instability conditions exist. If an air parcel at ground level is or quickly becomes warmer than the surrounding air, its density decreases. The air becomes extremely buoyant and rises explosively upward, building large cloud masses in the process.

Conditional instability exists when the lapse rate lies in between the wet and dry adiabatic lapse rates. The air near the surface is normally at the same temperature as the surrounding air. If this air is forced to rise, it cools at the dry adiabatic lapse rate. The dew point temperature is reached at some altitude, after which the air cools more slowly at the wet adiabatic lapse rate. It now becomes warmer at every successive altitude than the surrounding air and proceeds to rise by virtue of its buoyancy (lower density), building clouds in the process. There are other intermediate states that may exist in the earth's atmosphere but these three states generally summarize the typical tropospheric states on earth.

The lapse rates on the earth, as well as on other planets, may be termed adiabatic, sub-adiabatic, or super-adiabatic. Sub-adiabatic conditions would indicate a very stable atmosphere with limited motion. A super-adiabatic state would represent an atmosphere where atmospheric gases rose turbulently, i.e. a strongly convective situation. Adiabatic conditions would represent intermediate states of that planet's atmosphere where observations should note regions of stability and convective activity, dependent on local temperature variations. It is generally true that most changes occur in the troposphere on earth. This also appears to be true for the other planetary atmospheres although we may only judge what is happening by what we observe at the tops of the cloud layers.

The temperature increase in the stratosphere, as we return to a discussion of the earth's vertical temperature structure, is caused by the absorption of ultraviolet energy by molecular oxygen. The molecular oxygen photodissociates into atomic oxygen (O).

$$O_2 + \text{energy} \rightarrow O + O$$

The atomic oxygen combines with available molecular oxygen to produce ozone (O_3), another absorber of ultraviolet energy.

$$O + O_2 + M \rightarrow O_3 + M$$

M is any molecule which interacts with the molecular and atomic oxygen by absorbing some of the excess chemical energy of the reaction. Obviously, atmospheric photodissociation only occurs in the presence of sunlight. As a result, these reactions experience diurnal (daily) cycles with maximum concentrations of the end-products occurring during or immediately following the period of most intense sunlight. Minimum concentrations occur at night. The absorption of ultraviolet energy raises all atoms and molecules involved in a reaction to higher energy levels, i.e. a more excited state. A temperature sensor introduced into the stratosphere records more random collisions by these more energetic atoms and molecules and records an increase in temperature. The vertical temperature curve swings toward the right of the scale in the stratosphere in Fig. 1.2. Thus, the stratospheric temperature increase is due to the efficient absorption of ultraviolet energy. The by-products of the photochemical reactions are the ozone layer and a vertical temperature increase with increasing altitude in the stratosphere.

The interception of large quantities of ultraviolet energy by the stratosphere also plays a significant role in protecting life forms on the earth. It is assumed our early primordial atmosphere contained hydrogen (H_2), ammonia (NH_3), and methane (CH_4), later followed by an atmosphere of carbon dioxide (CO_2) and nitrogen (N_2). Oxygen was released to the atmosphere some time after the evolution of the first photosynthetic organisms. The initial oxygen released into the atmosphere was probably depleted rapidly by oxidation reactions with the lithosphere. Oxygen probably began to accumulate in the atmosphere after the demand for it in lithospheric oxidation processes lessened. Oxygen levels in the atmosphere increased and oxygen was dissociated to produce ozone. The protective ultraviolet shield we call the ozone layer was initiated at this time. Biologically, the reduction in the intensity of ultraviolet energy at the surface of the earth, especially the hydrosphere, permitted the evolution of additional photosynthetic species and the release of more oxygen into the atmosphere.

The mesosphere is the third major layer of the earth's atmosphere. The temperature decreases with height through the mesosphere. This is the normal trend we would expect to occur in all layers but only occurs in the troposphere and the mesosphere.

The temperature increase in the thermosphere (Fig. 1.2) is somewhat misleading. The density of the atmosphere at these altitudes is so low that the temperature increases recorded are merely the temperature sensor's response to occasional collisions by a very few, but highly energized particles. The temperature increases but the heat content of the atmosphere is negligible at these altitudes.

Wherever the vertical temperature profile does not decrease with height, we term the profile inverted. Hence an inversion, i.e. a reversal of the expected trend, characterizes the stratosphere and thermosphere. Inversion layers generally result from two processes. They may be caused by the absorption of solar energy by atmospheric gases. The gas molecules are raised to a higher energy state resulting in an increase in air temperature with height. Temperature inversions may also result from subsiding air, warming at the dry adiabatic rate, as it descends to lower altitudes. Regions of atmospheric subsidence are generally cloud free. The sinking air warms, preventing the condensation of water vapor into clouds. Clouds that may exist near the surface are often reevaporated or severly limited in the vertical thickness to which they may develop.

The zones between each of the major atmospheric layers are known as the pauses, i.e. tropopause, stratopause, and mesopause. Their temperature profile reveals a relatively constant (isothermal) temperature across each of the pauses. The pauses represent transition zones separating the thermal regime of one layer from that of the adjacent layer.

The heterosphere (Fig. 1.1) is a region from 85-700 km composed of ions. It virtually coincides with the earth's ionosphere (65-100 km). Ions are charged particles produced by intense solar radiation. The absorption of the solar energy raises atoms to an excited state. In that excited state, the atoms eject electrons, producing charged particles. Thus, the heterosphere consists of an abundance of free electrons and the positively charged atoms (ions) to which they (the electrons) were once attached. The rate of ion production is dependent upon the density of available atoms and molecules and the intensity of the incoming solar energy. The density of molecules and atoms decreases with height (Fig. 1.2), while the intensity of incoming solar radiation increases toward space. As a result, the maximum zone of free electrons is found at an intermediate altitude of 350 km. Localized electron concentrations within the heterosphere are responsible for the identification of the D, E, and F layers (Fig. 1.1), pronounced reflectors of radio wave transmissions. The total accumulation of ionized particles between 65-100 km is termed the earth's ionosphere. The ionosphere does overlap slightly into the homosphere. As previously mentioned, the ionosphere reflects radio signals, some regions more selectively than others. This property is used to propagate radio waves around the curvature of the earth providing man with global transmission capabilities. Needless to say, the advent of geo-stationary communications satellites has lessened our dependence upon the ionosphere's reflective characteristics. They (the communications satellites) also make our radio communications less vulnerable to disturbances in the structure of the ionosphere which occur during periods of intense sunspot and solar flare activity.

The structure of the earth's atmosphere provides a series of varied properties and characteristics which are generally understood at the macroscale level. Mesoscale and microscale processes are still under investigation. Recent evidence has revealed mesoscale interactions between layers of the atmosphere that had, heretofore, gone undetected or unnoticed. An example of this is the ozone-aerosol controversy. It was generally believed that mesoscale circulations between the upper troposphere and stratosphere either did not exist or were limited temporally and spatially. The proof of these interactions accompanied the discovery and identification of chlorofluorocarbon aerosol can propellents, which had been released at ground level, at stratospheric levels. This single example suggests that far more research is needed before we can safely say we fully understand the interactions between the four layers of our planet's atmosphere.

The Earth's Magnetic Field and the Magnetosphere

The earth's liquid metallic core conducts powerful electric currents which generate a magnetic field around the planet. Hence, the earth is a large magnet with north and south magnetic poles. The earth's magnetic poles are located relatively close to the geographic poles, however, they should not be viewed as stationary entities. The magnetic poles can and do shift their location in response to changes deep within the earth's core. The magnetic field surrounding the earth is presented in Fig. 1.3. The area within the limits of the earth's magnetic field is termed the magnetosphere.

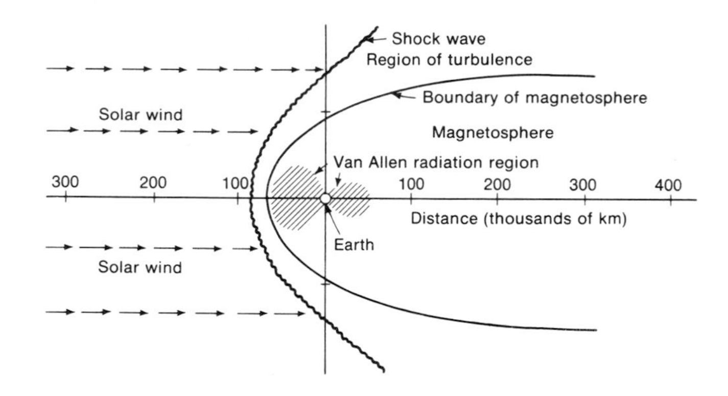

Fig. 1.3. The earth's magnetosphere. (Courtesy Miller & Thompson, Meteorology, 3rd Ed., Charles E. Merrill Publ. Co., 1979, p. 14.)

The outer boundary of the magnetosphere is termed the magnetopause. The magnetopause and the magnetosphere may be deformed due to interactions with the solar wind. The solar wind is the stream of ionized gas moving away from the sun at velocities >400 km sec^{-1}. Its composition is primarily hydrogen (95%) and helium (5%). The effect of the solar wind is to flatten out and compress the magnetosphere on the sun side of the earth. The region sunward of the magnetopause is an extremely turbulent zone. The sunward side of the magnetosheath delimits the location of the bow shock. The bow shock represents the region where the solar wind begins to deflect around the earth. The velocity of the solar wind decreases as the plasma temperatures increase. The sun facing side of the magnetosphere is limited to a radius of 10 earth radii (R_E) or 65,000 km by the strength of the magnetic field. On the opposite side of the earth, the magnetosphere is deformed into a tail that extends out away from the earth at least 1,000 R_E or 6,500,000 km.

Two zones within the magnetosphere trap energized particles from the sun. The lines of the magnetic force field steer and concentrate these particles into specific regions around the earth (Fig. 1.4). These regions are known as the Van Allen Radiation Belts. There are two Van Allen Belts. They are found at 2,600 km and 13,000-19,000 km from the earth's surface. The radiation in the outer belt is much more intense than in the inner belt. In general, the overall intensity of the Van Allen Belts varies with the degree of solar

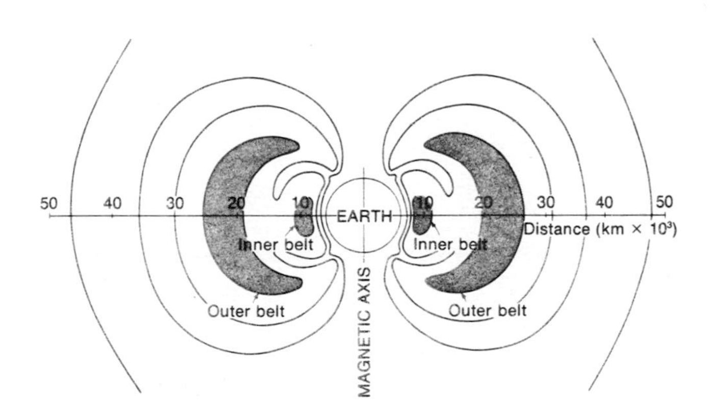

Fig. 1.4. The Van Allen radiation belts. The regions of maximum intensity are shaded. (Courtesy Miller & Thompson, Meteorology, 3rd Ed., Charles E. Merrill Publ. Co., 1979, p. 15.)

activity. Major solar disturbances, such as solar flares, eject intense ion streams into the earth's magnetosphere. The intensity of the trapped particles and hence, radiation, increases dramatically during these solar flare episodes. Occasionally, particularly energetic particles stream into and through the magnetosphere from the sun side passing over the magnetic poles into the upper atmospheric layers on the night side of the earth. Air molecules are ionized by collisions with these fast moving particles. In some atoms, the electrons are forced to move out away from the nucleus into a more unstable configuration, i.e. a higher electron shell, or stripped from the atom altogether. The electrons forced into the higher shell orbits will quickly return to their normal position (a lower shell), emitting a tiny burst of light energy. The spectacular displays of the northern lights (aurora borealis) and southern lights (aurora australis) occur when many electrons return to their lower energy states simultaneously. Auroral displays normally occur after local midnight when the earth-sun geometry is most favorable (Fig. 1.5).

Fig. 1.5. Photograph of the northern lights (aurora borealis) taken on May 3, 1978 in Sterling, Mass. (5 min. exposure). Photograph by J.P. Barbato.

The Earth's Energy Budget

The Nature of Energy

Energy may be transmitted in three ways: conduction, convection, and radiation. Conduction involves a molecule to molecule transfer of energy, i.e. a medium, such as the transferral of heat up the handle of a spoon sitting in a cup of hot coffee. Convection also involves a molecule to molecule transfer of energy but the medium's density (usually air or water in our discussions) changes and it flows. An example of this would be the air coming into contact with the forced hot water, baseboard heating elements in many homes. The air molecules in contact with the aluminum fins are raised to a higher energy state. The air's density changes and the air rises establishing a convection current. Radiation is the transfer of energy across a void, i.e. the sun's energy reaches us through space, which, for all practical purposes, is empty. Radiative energy transfer is accomplished by wave energy. The various types of wave energy oscillate up and down producing a pattern of peaks and troughs similar to ocean waves as the energy moves out away from its source. Forms of wave energy forms may be classified according to their wavelength. Wavelength is defined as the linear distance between two successive peaks and/or two successive troughs. The unit used to measure wavelengths of energy involved in atmospheric studies is the micrometer (μm), which is equal to 10^{-4} cm (0.0001 cm). The electromagnetic spectrum, which is a continuum of various wave energy forms, ranges from gamma rays, energy with the shortest wavelengths, to the longest wave forms, radio waves (Fig. 1.6). Sunlight consists primarily of ultraviolet (0.2 to 0.4μm), visible light (0.4 to 0.7μm) and infrared heat energy (0.7 to 80μm). The ultraviolet and the visible portions of sunlight are short wave energy often referred to as sensible or radiant heat.

The temperature of an object is the sole determinant of the forms of wave energy it will emit. Objects with high temperatures will emit more intense, short wave forms of energy. Objects with lower temperatures will emit energy more toward the longer wavelengths. The surface temperature of the sun is $6,000^{o}$K, while temperatures on earth range from 250°-300°K. Energy output curves for these temperatures are plotted in Fig. 1.7. The sun's energy output spans primarily the ultraviolet to the infrared with a peak output in the

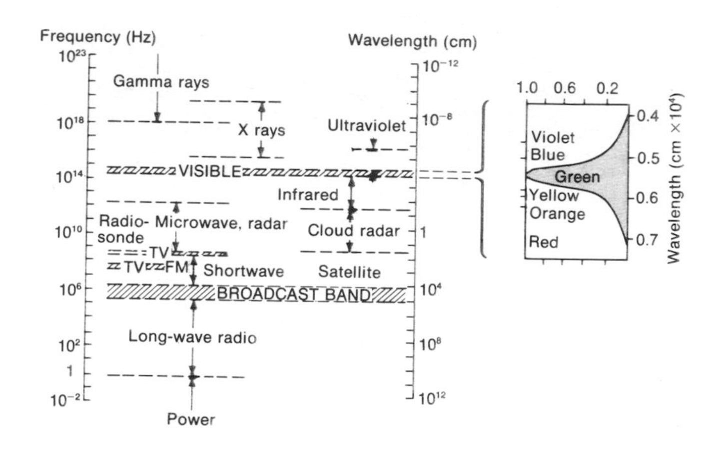

Fig. 1.6. The electromagnetic spectrum. Visible wavelengths are expanded at the right. (Courtesy Miller & Thompson, Meteorology, 3rd Ed., Charles E. Merrill Publ. Co., 1979, p. 63.)

visible wavelengths (0.4 to 0.7μm). The earth's atmosphere transmits much of the solar beam to ground level. There it is absorbed and reradiated as radiant heat energy. The earth and the lower atmosphere warm, the former from the absorption of short wave energy, the latter from the absorption of long wave energy. The ambient temperatures reached on the earth's surface and the oceans are much lower than the sun. The resulting energy emissions are in the form of long wave, infrared energy. The earth's infrared emissions span the band from 0.7 to 80μm with a maximum energy output in the 8 to 12μm range. The 8 to 12μm band is known as the radiation window through which the earth radiates its energy back to space.

The radiation window represents one of the most important balances between the gaseous composition of the earth's atmosphere and the reradiation of energy to space via the atmosphere. It has already been stated that the earth absorbs incoming short wave solar energy and reradiates long wave infrared energy. The earth must return every calorie of energy it receives from the sun back to space. During some geologic time periods, some of this solar energy has been stored, temporarily, as chemical energy in great coal beds. In general though, the earth must radiate the energy it absorbs back to space and the bulk of the reradiated energy is in the 8 to 12μm

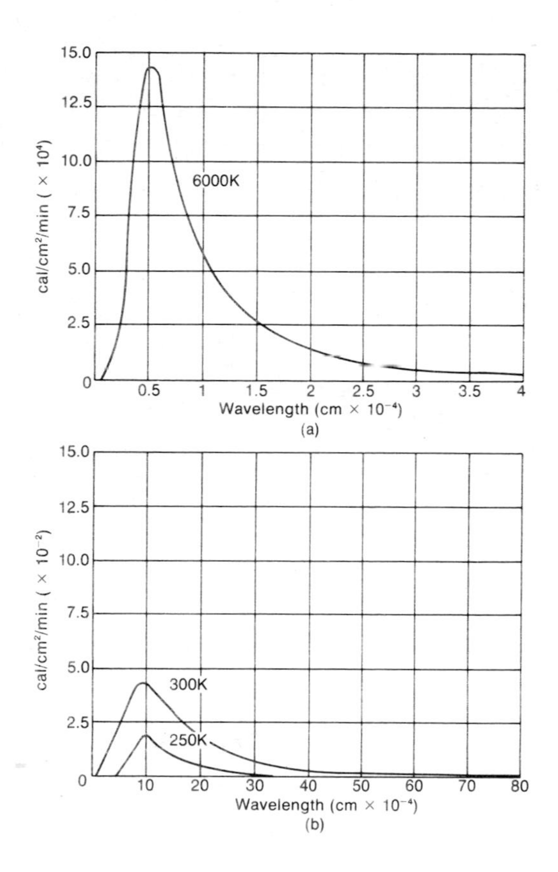

Fig. 1.7. Energy emission curves for the sun (6000°K) and the earth (250°-300°K). (Courtesy Miller & Thompson, Meteorology, 3rd Ed., Charles E. Merrill Publ. Co., 1979, p. 64.)

range (Fig. 1.6).

The variable atmospheric constituents, i.e. oxygen/ozone, carbon dioxide, and water vapor absorb long wave infrared energy very effeciently (Fig. 1.8). This absorption of heat energy explains how the lower atmosphere is heated. A closer examination of the combined absorption curves reveals that these three gases, four if you separate oxygen and ozone, barely absorb any infrared energy between 8 and 12μm. Thus the earth is free to radiate the bulk of its emissions through the atmosphere and into space. This process is termed radiational cooling.

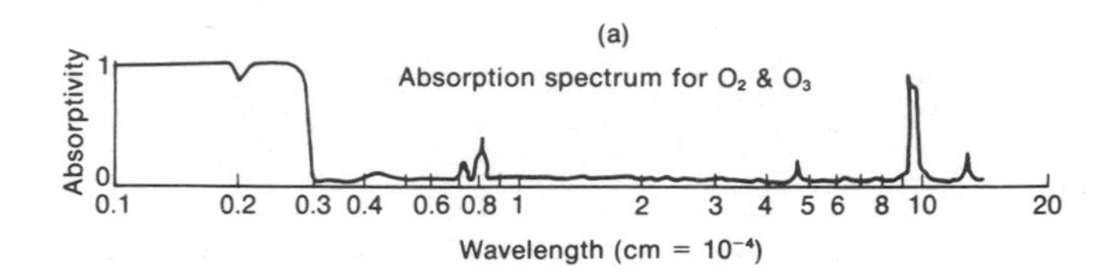

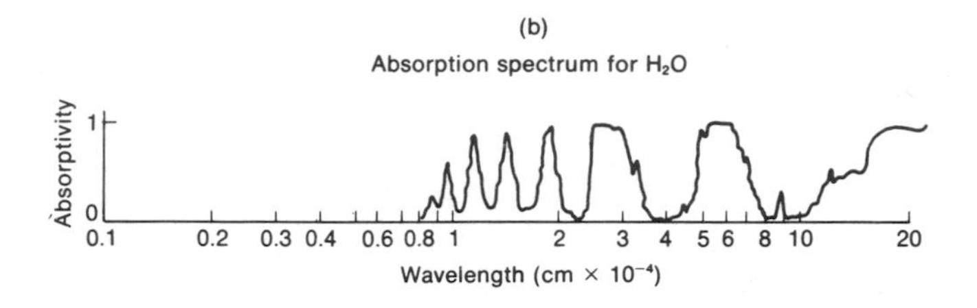

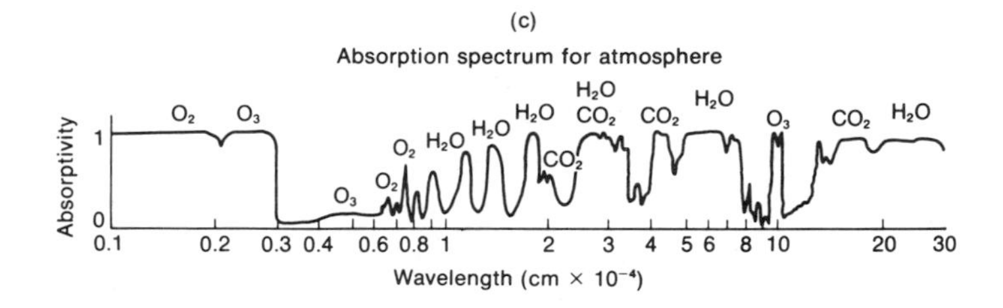

Fig. 1.8. Absorption of electromagnetic radiation at various wavelengths by (a) O_2 & O_3; (b) H_2O; and (c) the primary absorbing gases of the atmosphere. (Courtesy Miller & Thompson, Meteorology, 3rd Ed., Charles E. Merrill Publ. Co., 1979, p. 66.)

Radiational cooling involves the emission of infrared energy by a surface. The soil surface and sea level provide the necessary surfaces on earth. Radiational cooling proceeds on cloudless nights with very gentle winds. The clear skies are a requirement because cloud layers consist of water droplets around condensation nuclei, usually dust or sea salt particles. The clouds absorb the infrared energy and reradiate it in all directions including back down to the earth. There it is reabsorbed and reradiated. The net result is that clouds act similar to an insulating blanket and thus interrupt and reduce the flow of infrared energy from the ground level to space. The gentle winds help to mix the atmosphere next to the ground and create turbulence. Radiational cooling creates a layer of colder air near the ground because this layer has given up its heat energy to the ground. The ground, in turn, radiates that heat energy out to space. The air a few meters above the ground is slightly warmer. The eddy motions (turbulence) generated by the light winds mix and transport the warmer air aloft down to ground level. The warmer air now in contact with the ground, radiates its heat energy to the earth which in turn radiates it out to space. In this manner, the lower layer of the troposphere is cooled by radiational cooling.

Radiational cooling proceeds in a nightly as well as seasonal pattern. Those portions of the earth that experience radiational cooling on consecutive nights begin to cool down. Radiational cooling is more rapid during the fall and winter months because the nights are longer and less solar energy is received per unit area of the surface during the daylight hours. Radiational cooling is therefore responsible for diurnal nighttime cooling as well as the seasonal hemispheric cooling of the earth.

The ability of a planet to return its absorbed solar energy to space determines many factors such as temperature regimes, storm development, and duration. The role of radiational cooling's effect on temperature regimes is clear from the preceeding discussion. The role of storms as another process which returns heat energy to space will be explored in a later section of this chapter. The duration of certain storms is determined by the need to exchange energy. On earth, energy exchanges are rapid and storms normally last only a week or two. Our study of Jupiter will reveal a cloud shrouded planet so slow to cool that certain "storms" may exist for over three centuries.

The Energy Budget

The sun emits a vast amount of energy but the earth only intercepts a miniscule amount. The amount received at the outer edge of the earth's atmosphere is ≈2.0 g cal cm^{-2} min^{-1}. Since 1.0 g cal cm^{-2} min^{-1} is equal to 1.0 langley (ly), the amount of solar energy reaching the earth's outer periphery is 2.0 langleys and is termed the solar constant. Once this energy penetrates the atmosphere, it is reflected, diffused, and absorbed as part of the heating and cooling processes of the earth, collectively known as the energy budget. The energy budget of the earth is itemized in Table 1.3 and diagrammed in Fig. 1.9.

TABLE 1.3

Energy budget of the earth's atmosphere.
(After Miller and Thompson, Meteorology, 3rd ed., courtesy of Charles E. Merrill Publishing Company, 1979, page 72.)

	Langleys	
I. INCOME—Solar radiation		
Average solar radiation received by earth and atmosphere:		0.50
Absorbed by:		
(a) atmosphere and clouds	0.10	
(b) earth's surface	0.25	
Net short-wave absorbed	0.35	
II. OUTGO—Short-wave radiation		
Lost to space by:		
(a) reflection from clouds	0.10	
(b) reflection from earth	0.02	
(c) scattering from air	0.03	
(i) Net short-wave outgo	0.15	
III. OUTGO		
The atmosphere:		
(a) radiates to the earth	0.53	
(b) radiates to space	0.32	
(c) receives from earth through radiation	-0.57*	
(d) receives from earth through convection, turbulence, and evaporation	-0.15	
(ii) Net outgo from atmosphere	0.13	
The earth:		
(a) radiates to space	0.03	
(b) radiates to the atmosphere	0.57	
(c) loses to the atmosphere through convection and turbulence	0.04	
(d) loses to the atmosphere through evaporation	0.11	
(e) receives from the atmosphere through radiation	-0.53	
(iii) Net outgo from earth	0.22	
IV. TOTAL NET OUTGO (i+ii+iii)		0.50

*Negative signs under "outgo" indicate net gains.

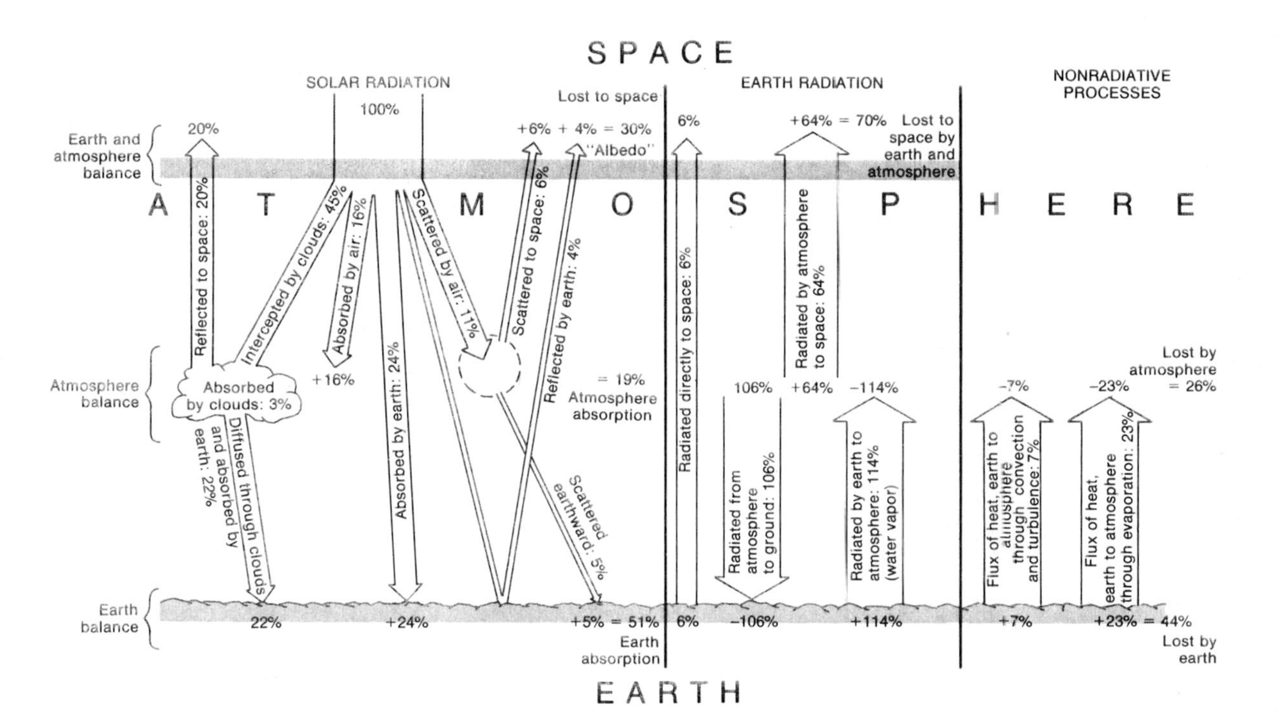

Fig. 1.9. The earth's energy balance. (Courtesy Miller & Thompson, Meteorology, 3rd Ed., Charles E. Merrill Publ. Co., 1979, p. 71.)

The earth's energy balance (Fig. 1.9) may be divided into the short wave side and the long wave side. The short wave side of the energy budget explains what occurs as solar radiation penetrates into the earth's atmosphere. The long wave side of the energy budget explains how the earth reradiates the energy it has absorbed. The long wave side of the energy budget involves both radiative and non-radiative processes.

The solar radiation reaching the outer periphery of the earth's atmosphere is assumed to be 100 percent or unity (1.0) (Fig. 1.9). Forty-five percent of the incoming short wave energy, primarily ultraviolet and visible light, is intercepted by the clouds. The clouds distribute this energy by immediately reflecting 20 percent back to space, absorbing 3 percent, and diffusing 22 percent down where it is absorbed by the earth's surface. Sixteen percent of the incoming solar beam is absorbed by the atmosphere, the bulk of it by the oxygen and ozone in the stratosphere. Approximately 24 percent of the solar beam passes directly through the atmosphere as direct solar radiation and is absorbed by the ocean and land surfaces. Four percent is immediately reflected to space by the ground and ocean surfaces. Scattering by air molecules (11%) returns 6 percent to space and 5 percent down to the earth's surface where it is also absorbed. The scattering of the blue wavelengths is responsible for the blue color of our atmosphere. The energy immediately reflected to space (top left, Fig. 1.9) is 20% + 6% + 4% = 30%. The 30 percent total of energy reflected to space represents the earth's albedo. The albedo may also be noted by stating that 0.3 of all incoming sunlight is immediately reflected to space. The total amount of energy absorbed by the earth's land and ocean surfaces (bottom left, Fig. 1.9) is 22% + 25% + 5% = 51%. The earth absorbs slightly over one-half of the solar constant (or ≈1.0 ly). The atmosphere (middle left, Fig. 1.9), including clouds, absorbs 16% + 3% = 19%. The albedo (30%), the atmospheric absorption (19%), and the surface absorption (51%) values total 100 percent. Therefore, the energy budget has accounted for all incoming solar (short wave) energy.

The long wave side of the energy budget explains how the earth reradiates the short wave energy it has absorbed (51% + 19%). Three processes govern the return of energy to space: direct infrared radiation, convection, and latent heat transfer. Direct infrared radiation (6%) proceeds through the radiation window at certain wavelengths (8 to 12μm) through the process of radiational cooling. Not all infrared energy passes directly out to space. Some infrared energy radiates from the earth and ocean surfaces to the atmosphere. There it is absorbed and reradiated, some going to space, some back down toward the earth. The net loss (114-106%) is 8 percent. Therefore, infrared reradiation is responsible for removing 14 percent (6% + 8%) of the absorbed energy from the earth's surface to space.

Convection and turbulence remove 7 percent of the infrared energy from the earth's surface. Convective clouds, particularly thunderstorms play an important role in redistributing energy from ground levels into the upper troposphere.

The most significant energy exchange process on earth is the transfer of latent heat energy (23%). When water is evaporated from the surface of the oceans, lakes, etc., approximately 540-600 calories per gram of water are needed to convert liquid water to the vapor state. This heat energy does not warm the air, it merely keeps the water in the vapor phase. The processes or orographic uplift caused by mountain ranges, strong convection (thunderstorms), and cyclonic storms (low pressure systems) all act in their own unique manner to force the moist air to rise. As it rises, it cools. If the air cools to the dew point temperature, it becomes saturated and moisture begins to condense out as cloud and possibly rain droplets. At the moment of condensation, 540-600 calories (for every gram of water) of latent heat are released to the atmosphere as sensible heat. Its journey began when the energy was transferred by evaporation from the warm tropical and subtropical oceans to the atmosphere. Meridional flows transported the energy into the mid-latitudes where it was then released into the atmosphere in association with condensation processes. Once in the upper atmosphere, this energy radiates out to space completing the long wave side of the energy budget.

The long wave side of the energy budget (Fig. 1.9) may be summarized by considering the various pathways energy has followed from the earth and the atmosphere out to space. The earth experiences an energy loss amounting to 44 percent (bottom right, Fig. 1.9), i.e. 6% + (114-106 = 8%) + 7% + 23%. Thirty-eight percent is lost to the atmosphere while 6 percent was lost directly to space (38% + 6% = 44%). The atmosphere (middle right, Fig. 1.9) experiences losses (over and above the 44% from the earth) of 26 percent (106% + 64% - 114% - 7% - 23%). The total amount lost by the atmosphere totals 64 percent. Total earth losses (top right, Fig. 1.9) equal 70 percent (6% + 64%). The budget is balanced as the earth absorbed 70 percent of the incoming solar radiation (51% + 19%) and reradiated, as infrared energy, 70 percent (64% + 6%). It is clearly evident that long wave energy absorption by carbon dioxide, water vapor, and particulates play an important role in modulating the earth's energy budget.

The energy budget of the earth is often compared to the solar heating of a greenhouse. The incoming short wave energy (both ultraviolet and visible light) penetrates the atmosphere. The energy is absorbed, scattered, reflected back to space, and some fraction of the original amount reaches the ground. The total amount reflected back to space (30%) is termed the earth's albedo. In the "greenhouse effect," the incoming energy is also scattered, reflected, and absorbed. The glass absorbs the ultraviolet energy, whereas, in the atmosphere it is absorbed by the ozone layer. The visible wavelengths penetrate to the inside of the greenhouse, where they are absorbed by the soil and plants. The soil and plants are warmed and radiate at infrared wavelengths. Since glass does not allow infrared energy to freely penetrate, the energy is trapped inside the greenhouse and the air warms. In the environment, the earth and ocean surfaces are heated and in turn radiate in the infrared wavelengths. The infrared energy is absorbed by water vapor, carbon dioxide, and particulates in the lower atmosphere and hence, the air warms. The solar heating of an atmosphere and a greenhouse are analogous situations. This explains the use of the term, "greenhouse effect," in reference to the heating of a planetary atmosphere.

Man's technology has also added tons of carbon dioxide and other gaseous by-products to the atmosphere. Rising carbon dioxide levels are particularly worrisome because of the dominant role the gas plays in the absorption of infrared energy. Will continued increases in carbon dioxide alter the energy budget and create irreversible climatic changes? It is probably impossible to say at this point in time but let us hope that man will pay particular attention to the consequenses of his own role in atmospheric modification.

Atmospheric Pressure

The vertical pressure decrease with increasing altitude was presented earlier in Fig. 1.2. The decrease of pressure with height is linear, dropping 1/30 of the total for each 275 m increase in altitude. The global, horizontal pressure distribution is a function of the differential heating of the earth's surface by the sun.

The mean annual values for the amount of solar energy absorbed by the earth as compared to the heat emitted by the earth is presented in Fig. 1.10. The surface absorption curve is flattened in the tropical latitudes indicating the uniformly high absorption of solar energy. The absorption curve dramatically decreases poleward demonstrating the lesser amounts of energy received in the polar zones.

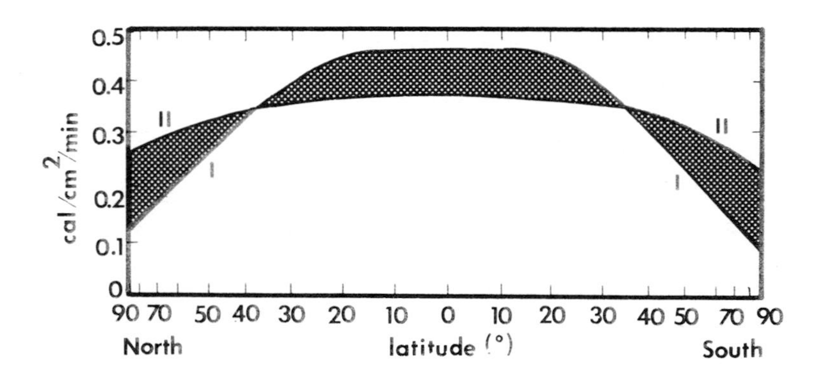

Fig. 1.10. Mean annual radiation absorbed (I) and emitted (II) by the earth and its atmosphere. (Courtesy Miller & Thompson, Meteorology, 3rd Ed., Charles E. Merrill Publ. Co., 1979, p. 78.)

The amount of solar energy received in the various latitudinal belts as well as in the polar regions is related to the axial tilt, the parallelism of the axis, and orbital distances from the sun. The earth's axis is tilted 23.5^{o} from a perpendicular constructed to the plane of the ecliptic. The plane of the ecliptic is an imaginary geometric plane extending out from the sun in all directions. Most of the planets of our solar system have orbits that are aligned with, or intersect, the plane of the ecliptic. An imaginary line drawn along the earth's axis from the geographic south pole to the geographic north pole and then out into space will extend to the vicinity of Polaris, the North Star. The axial tilt and parallelism introduce an annual cycle in the orientation of the sun facing hemisphere. During the northern hemisphere's summer, the northern portion of the globe is tilted toward the sun. The sun's rays are more direct so the surface energy receipt per unit area is greatest at this time of year. Northern hemisphere summer is also the time when that half of the globe illuminated by the sun extends from and includes the entire north polar region to the Antarctic Circle (Fig. 1.11). The northern hemisphere is therefore in the daylight for a longer number of hours per day which also results in greater energy absorption.

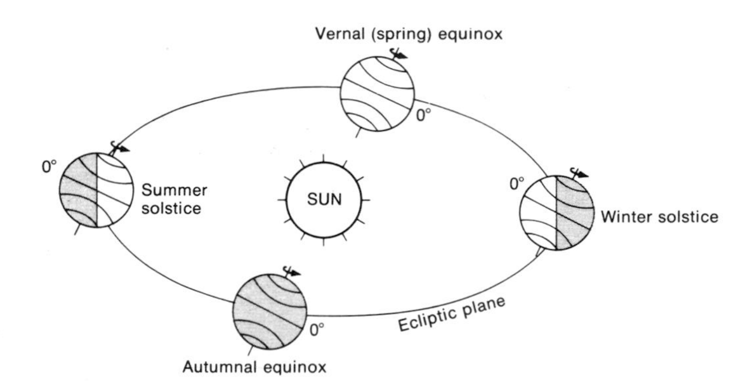

Fig. 1.11. Earth - sun orbital relationships. (Courtesy Miller & Thompson, Meteorology, 3rd Ed., Charles E. Merrill Publ. Co., 1979, p. 76.)

The earth reaches aphelion during the first week in July, several weeks after the summer solstice. Since the earth is farthest away from the sun, it receives 3 percent less solar energy. The northern hemisphere is thus deprived of this energy, however, the amount is so small as to be negligible in relation to the earth's total energy budget. The general trend in Fig. 1.10 shows a significant absorption of energy in the tropical and subtropical latitudes decreasing from the mid-latitudes poleward.

The heat emitted curve (Fig. 1.10) shows that more energy is emitted than is absorbed by the polar regions. It is clearly evident that there must be a transfer of energy from the tropical latitudes poleward since the tropics and subtropics radiate less energy than they have absorbed. This energy exchange is referred to as latent heat transfer and involves both the transfer of the latent heat of evaporation by the global wind systems and the poleward flow of warm, ocean currents.

The Nature of Winds

Macroscale Winds

Temperature differences produce pressure differences. Since atmospheric pressure differences are responsible for the initiation of winds, let us first examine the nature of winds before examining the global circulation of the earth.

The movement of the atmosphere results from differences in pressure, either locally (mesoscale) or regionally (macroscale). Macroscale pressure differences occur over hundreds of kilometers whereas mesoscale differences are generally restricted to 0-100 km in extent. The overall rule of air pressure relationships is that air always flows from regions of higher pressure to regions of lower pressure. Wind speeds are determined by the magnitude of the pressure difference, i.e. larger differences result in faster winds. Surface winds on earth actually are the end product of the interaction of three "forces": the pressure gradient, the Coriolis effect (force), and friction (Fig. 1.12). The arrows in Fig. 1.12 are vectors which represent the direction of each force. The pressure gradient force initiates the movement of air in a direction toward the center of a low pressure area. The moment the air begins to flow, it is deflected to the right of its path by the Coriolis effect (force) in the northern hemisphere (to the left of its path in the southern hemisphere). The Coriolis effect is an apparent force, not a real force. The air is not really flowing in a curved trajectory to the right of the pressure gradient. What actually occurs is that the air flows at a particular speed in a particular direction but, all the while, the earth is rotating underneath the atmosphere. As observers fixed to the earth by gravity, we see a deflection or right hand turning from the original direction taken by the air.

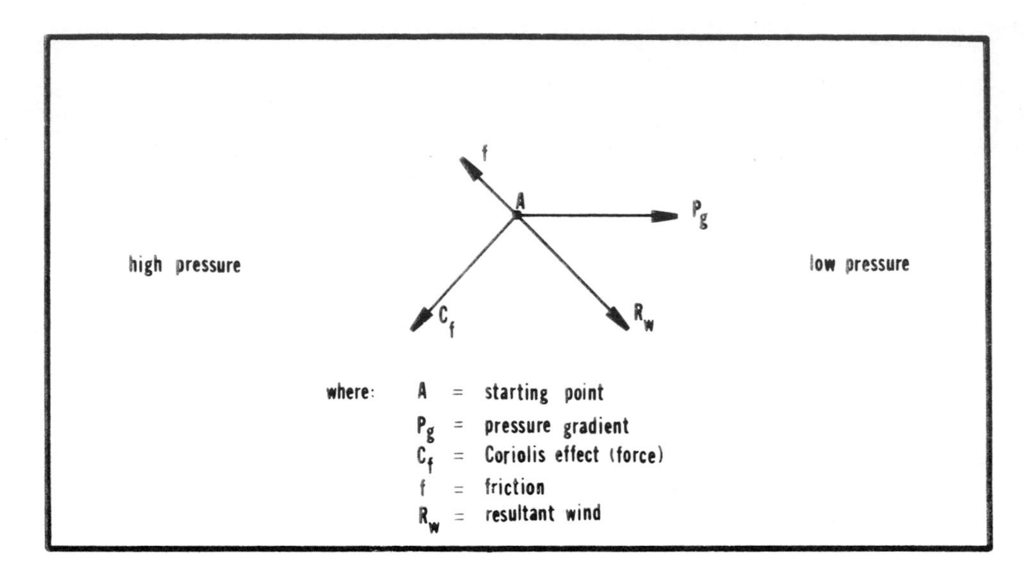

Fig. 1.12. Forces that initiate surface winds.

The air follows curved trajectories because the earth has turned on its axis underneath the atmosphere. This effect is often designated as the Coriolis "force," however, as previously stated, it is only an apparent force. The Coriolis effect on the wind is greater with increasing geographic latitude and is nonexistent at the equator. It is also partially dependent on wind velocity so the faster the wind flow the greater the Coriolis effect. The Coriolis effect is primarily responsible (along with friction) for air movement along curved paths often resulting in the formation of rotating weather systems.

The balance between the pressure gradient force and the Coriolis effect is termed the real or resultant wind (R_W). Friction between the land surface and the air slows the resultant wind. The rougher the surface terrain, the greater the frictional drag. Therefore, the friction vector is diagrammed in a direction opposite the resultant wind. Friction decreases rapidly with height and is all but nonexistent above 1,000 m.

High altitude winds in the troposphere are termed geostrophic winds (Fig. 1.13). The explanation of the forces remains the same but the friction is absent. High altitude winds are therefore faster

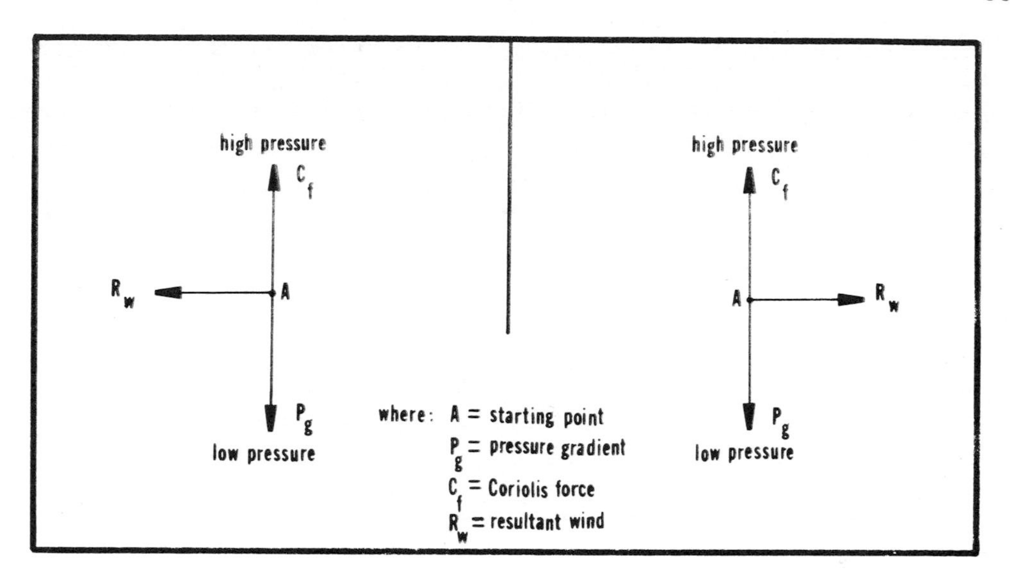

Fig. 1.13. Forces that initiate upper level geostrophic winds.

and tend to flow in zonal (east to west or west to east) paths.

The pattern of high altitude winds is not uniformly zonal on earth. Continents, because they heat and cool more rapidly, affect the atmospheric pressure patterns. The high altitude winds respond to these variations by deviating from their zonal course and bending equatorward over the land masses and poleward over ocean areas. The result is a series of U-shaped troughs or Rossby Waves. The fast flows associated with the Rossby Waves are referred to as jet streams. Since pressure differences are largest during the winter when the thermal contrasts across the continents are greatest, the geostrophic winds (and jet streams) exhibit their highest velocities during the winter months.

Wind Speeds

Wind velocities are dependent on the forces which initiate and act to modify both surface and geostrophic winds. The maximum surface winds on earth vary with the strength of the pressure gradient. Wind speeds in the trade wind belt routinely average between 10 km hr^{-1} and 19 km hr^{-1}. The westerly winds often average 16 km hr^{-1}. Organized storm systems, where latent heat transfer is occurring, may be accompanied by higher wind velocities. Mid-latitude cyclones often have 32-48 km hr^{-1} winds, while hurricanes (tropical cyclones) have winds in excess of 125 km hr^{-1}. The upper level jet streams commonly move around the earth at velocities of 160 km hr^{-1} and often reach velocities of 400-480 km hr^{-1} during January and February when the thermal contrasts during the northern hemisphere winter are at a maximum. It is difficult to generalize and categorize the topic of wind speed. The regional weather conditions determine wind speeds. The actual velocities will normally fall within the ranges given for each type of phenomena.

Mesoscale Winds

Mesoscale winds result from local differences in pressure produced by heating or radiational cooling of a surface. The most common mesoscale wind is the sea breeze (Fig. 1.14). The sea breeze

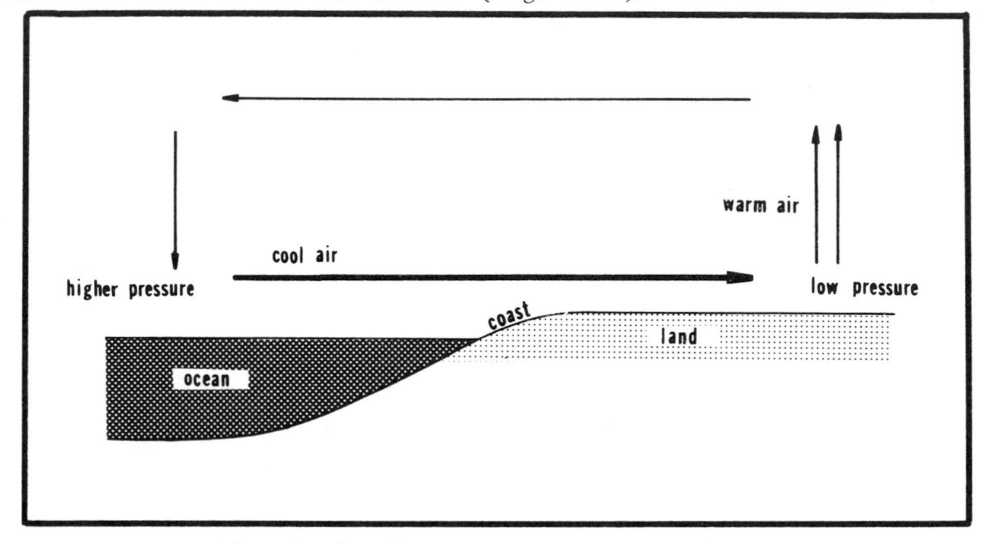

Fig. 1.14. The sea breeze circulation.

normally develops during the summer months when the temperature contrasts between the adjacent land-ocean surfaces are greatest. Sunrise initiates a rapid heating of the land surface. The water is much slower to heat since it distributes the heat energy to a greater depth and has such a very high thermal capacity. Generally, by 1000 hours (local time), the coast is significantly warmer as is the air above the land. The warm air rises creating a local region of lower barometric pressure. The air over the water is cooler and therefore denser, creating a local region of higher pressure. An air flow from higher to lower pressure begins, advecting some of the cooler air inland. The sea breeze may flow from any compass direction. The orientation of the coast and ocean determines the direction of flow but it will always be from the water toward the land, i.e. an onshore wind. In reality, sea breezes are more complex. More thorough discussions of the complexities and vagaries of sea breezes are available in current journal literature (see references).

Mountain and valley breezes are another form of local wind (Fig. 1.15). During the day, the valley air warms and moves upslope in response to the change in density, i.e. the valley breeze (Fig. 1.15A). Nighttime conditions, particularly in mountainous

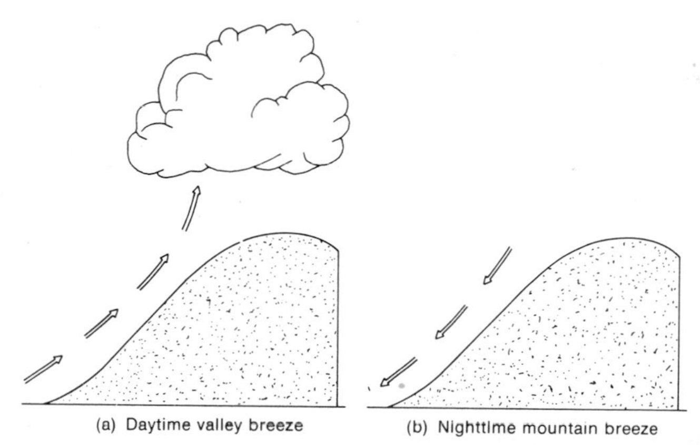

Fig. 1.15. Valley and mountain breezes. (Courtesy Miller & Thompson, Meteorology, 3rd Ed., Charles E. Merrill Publ. Co., 1979, p. 140.)

areas, favor rapid radiational cooling, as there is less water vapor in the air and generally fewer clouds to interfere with the cooling process. The air adjacent to the mountain slopes cools and flows downhill in response to its greater density (Fig. 1.15B). Winds therefore tend to be downslope in mountainous regions during the night.

Similar local conditions prevail in plateau regions where a mass of cold air may be formed by radiational cooling. The cold air builds up and, because of its density, flows or "drains" downhill off the plateau (Fig. 1.16A). These drainage flows are called katabatic winds. Normally, their velocity is not >4 m sec^{-1}, however, where local topography restricts the flow, velocities may reach 40-60 m sec^{-1}.

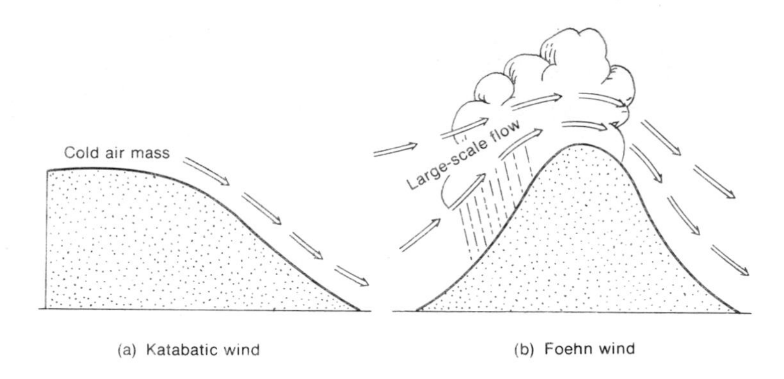

Fig. 1.16. Katabatic (drainage) (a) and foehn (b) winds. (Courtesy Miller & Thompson, Meteorology, 3rd Ed., Charles E. Merrill Publ. Co., 1979, p. 141.)

Foehn winds are intermediate examples because they are local winds but are induced by macroscale flows (Fig. 1.16B). The regional wind forces air to rise orographically over the windward slopes of mountains. Moisture and latent heat are released in amounts commensurate with the original moisture content of the air and the amount of forced vertical ascent. The descending air on the leeward side of the mountains warms adiabatically at the dry rate. Since the heat content of the air increased with the latent heat of condensation on the windward side, the descending flow warms rapidly. This flow is called the foehn or chinook wind and often results in the rapid melting of snowfields creating flooding and/or mudslides in the process.

Mesoscale winds occur in many locations on earth. All of the mesoscale winds result from pressure differences induced by thermal heating and/or radiational cooling of the earth's surface. These winds may only exist if the macroscale winds are very gentle. The foehn is the exception to this statement. High regional wind speeds obliterate the thermal differences and prevent the formation of mesoscale wind regimes.

Global Pressure Belts

The differential heating of the earth's surface creates seven major pressure belts across the earth (Fig. 1.17). The warmer air in the equatorial zone rises because it is less dense. It flows out

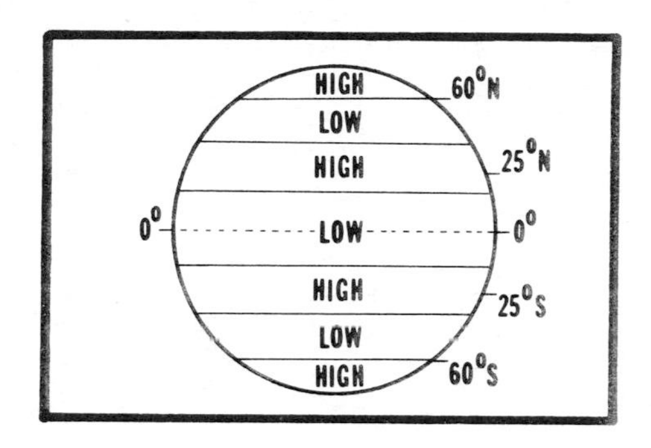

Fig. 1.17. The earth's major atmospheric pressure belts.

away from the equator at higher altitudes cooling radiatively to space. As the air cools, it becomes denser and sinks in the subtropics. Rising air lowers surface barometric pressure, hence the equatorial zone of low pressure. Sinking air increases surface barometric pressure creating a zone of high pressure between latitudes 10°-35°N and 10°-35°S. The sinking air in the subtropics bifurcates near the surface with some flowing equatorward and some flowing poleward. In the polar regions, radiational cooling creates the large, cold, high pressure systems within which air is sinking. The cold polar air flows equatorward meeting the warmer air flowing poleward from the subtropics. The warmer air is forced to rise up over the cold air, and condensation, clouds, and precipitation occur periodically in the mid-latitudes. The rising air also produces a belt of low pressure in the mid-latitudes. This interaction also generates fronts and rotating cyclonic storms.

The General Circulation of the Earth: The Tricellular Model

A description of the planetary wind system may be approached in two steps. The first is to describe what the winds would be like on a non-rotating earth. The second step would examine the role of the Coriolis effect on the general circulation of the earth. The wind flow on a non-rotating earth would exhibit one of two patterns. On a planet with a uniform surface, i.e. all water, the circulation would begin as air rose over the equator, flowed poleward, subsided over the poles and returned equatorward along the surface. A single "Hadley" cell would dominate each hemisphere with meridional flows (N↔S). The other pattern, more representative of the earth's land and water areas, would divide each hemisphere into three cells based on the subsidence of air in the subtropical latitudes due to rapid radiational cooling of the air aloft (Fig. 1.18). The flow patterns

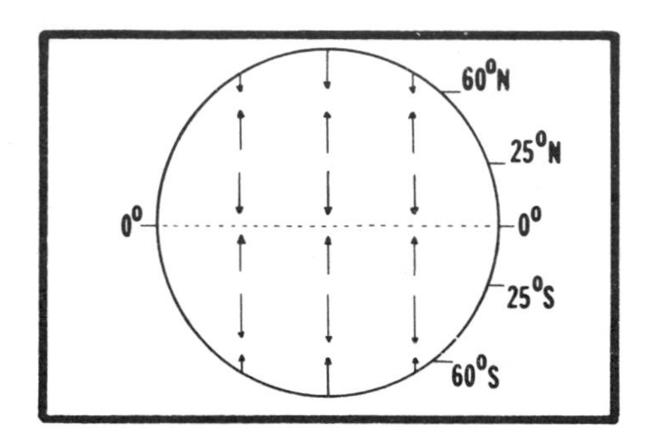

Fig. 1.18. Winds on a non-rotating earth.

would remain meridional.

If the non-rotating earth were completely covered with water, a composite map combining the wind and pressure belts (Figs. 1.17 and 1.18) would suffice to describe the basic pressure and wind relationships in the atmosphere. In fact, the earth does rotate from west to east and 25 percent of the earth's surface is land. The rotation creates the Coriolis effect. Land surfaces also exert an influence on the global wind patterns, particularly in the subtropics. The general circulation of the earth, subject to several limitations, would be represented by Fig. 1.19. The warmer, equatorial air would rise and flow poleward at higher levels in the troposphere, cooling radiatively to space all the while. The density of the air increases with decreasing temperature and subsidence begins at, or near, 25°N and 25°S. The subsiding air warms at the dry lapse rate and contributes to the warm conditions prevalent in the subtropics. The subsiding air increases barometric pressure resulting in a subtropical, semi-permanent belt of high pressure.

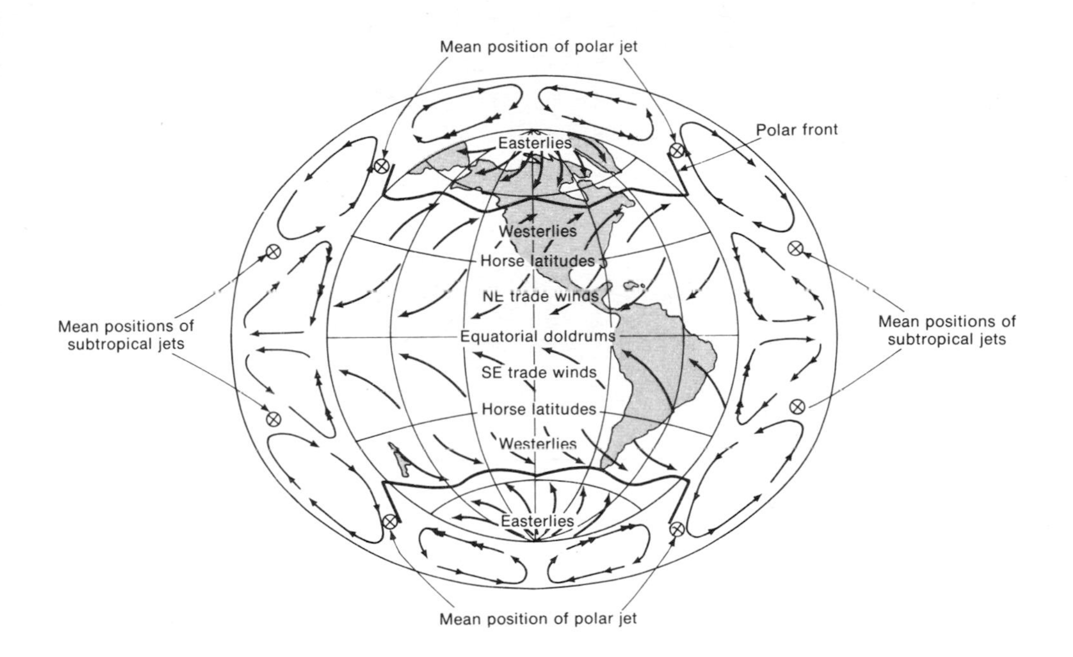

Fig. 1.19. The general circulation of the earth. Double-headed arrows indicate wind component from the east. (Courtesy Miller & Thompson, Meteorology, 3rd Ed., Charles E. Merrill Publ. Co., 1979, p. 123.)

The air cannot continue subsiding into the earth and oceans so it begins to flow both equatorward and into the mid-latitudes. The surface wind belts known as the northeast trade winds and the westerlies are the result of these flows. They are not meridional flows because the Coriolis deflection causes a deflection to the right of the wind path in the northern hemisphere (left, in the southern hemisphere). The polar latitudes are source regions for cold air masses produced by radiational cooling. This air moves out in an equatorward direction but soon deflects to the right of its path producing the polar easterly wind belt.

The convergence zone between the polar easterlies and the westerlies delimits the location of the polar front. Here, the warmer, moister air masses collide with the colder, drier air masses. The warm, moist, air is forced to rise up, over the colder air. At the same time, rotational motions generated by the air mass collisions induce the birth of an extratropical cyclone (mid-latitude low pressure system). The rising air reduces atmospheric pressure, releases vast quantities of latent heat energy, and completes the return flow circulations associated with the three-celled model in Fig. 1.19.

Several modifications of the tri-cellular model are necessary. The surface wind belts, depicted with the subsiding air and subtropical belt of high pressure, persist globally during the winter months. During the summer months, the subtropical belt of high pressure breaks down as the land surfaces heat and create localized areas of low pressure over the continents. The high pressure belt is transformed into two large semi-permanent oceanic high pressure systems. The winds and pressure associated with these high pressure systems would remain the same as depicted in Fig. 1.19.

The upper level winds are dominated by the westerly flow. The "antitrades," i.e. the flow aloft above the surface trade winds, are a gentle flow whose summertime core is marked by the easterly, subtropical jet stream (Fig. 1.19). The "antitrades" are often very weak or entirely absent. The westerlies often dominate both the surface and upper level circulation of both hemispheres. The core of the upper level westerlies is the westerly jet stream. Strong thermal gradients result in strong pressure gradients which produce high speed geostrophic winds. The westerly jet stream velocities often exceed 240 km hr^{-1}.

The three-celled model presents the general circulation of the earth reasonably well. The pattern is a relatively basic one consisting of three prevailing wind belts in each hemisphere. The general circulation of the earth is far more complicated than it may appear. The complexity of this system is underscored by the overall fact that there are probably more than 100 variables which, alone or in combination, may alter the growth and development of clouds, precipitation, and weather systems.

The Monsoons

The general circulation of the earth is a reasonably orderly pattern of winds and pressure belts. There are additional numerous variations in the global wind field, particularly at the mesoscale level, however, these are more important to students of the earth's meteorology. The student of planetary atmospheres need only concern himself/herself with local winds such as sea breezes, mountain and valley breezes, katabatic, and foehn (chinook) winds. There is one major perturbation in the general circulation of the earth that must be discussed. That is the wind system known as the monsoons of Southern and Southeast Asia (Fig. 1.20). During the winter months, the wind flow from South and Southeast Asia is from the northeast toward the equator (Fig. 1.20B). Some authors refer to this

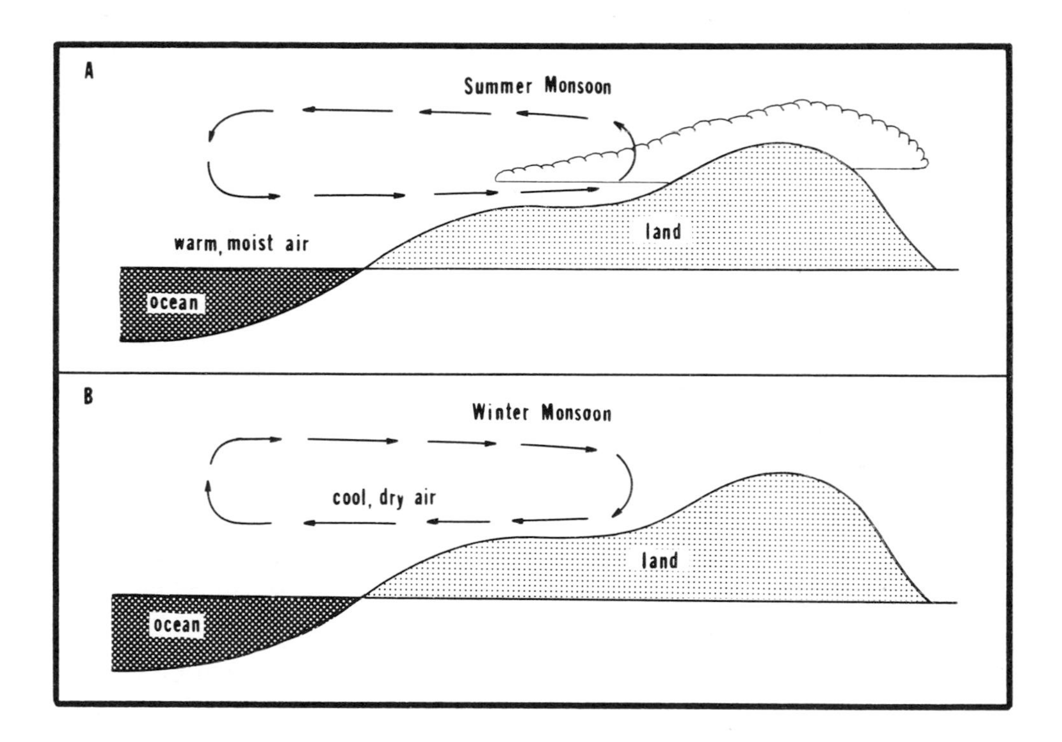

Fig. 1.20. The monsoon circulation.

northeast flow as the winter monsoon but this is somewhat of a misnomer. There is nothing unusual about the winter wind regime in this part of the world. It is normal and is part of the general circulation of the earth. It is the summer wind regime that is significant because it represents a regional reversal of the normal planetary wind field from northeast to southwest.

The summer wind reversal is a response to the superheating of the land in the Indus-Ganges River Valleys (i.e. the Indo-Gangetic Plains). The Himalaya Mountains to the north prevent any cold, dry Siberian air from reaching northern India. The result is that by late spring this region experiences daytime temperatures in excess of 318°K. The air above the land is warmed and rises, creating a region of lower atmospheric pressure (Fig. 1.20A). The rising air cannot flow poleward because of the Himalaya Mountains, so it flows out at elevations above 2 km toward the equator. Since the rising air must be replaced, an air flow is initiated from the southwest. There are a number of conditions that must be met before the onset of the southwest monsoon flow. The first is the superheating of the Indo-Gangetic Plains and the creation of a large, thermal, low pressure area with rising air. The high pressure belt that normally prevails from the Atlantic Ocean eastward to the Pacific Ocean at these latitudes must divide and recede eastward and westward. The subsidence associated with the belt of high pressure must weaken and eventually disappear entirely, permitting the warm air created by the thermal low to rise to higher elevations. The final obstacle to the southwest monsoon flow is the jet stream. The jet is associated with the polar front, the boundary between the cold, dry, polar, Siberian air and the warmer, moister, subtropical air. The Himalaya Mountains influence even the upper level wind flows because they are such a massive topographic obstacle. The westerly (polar) jet stream bifurcates with one branch flowing west to east, north of the Himalayas over the Tibetan Plateau and the other branch flowing west to east, south of the Himalayas. The presence of the jet stream contributes to the stability of the atmosphere above the Indo-Gangetic Plains. However, the polar front recedes poleward as spring progresses and southern Siberia begins to warm from increased solar heating. The northward retreat of the polar front marks the disappearance of the jet stream south of the Himalayas. Conditions for the warm air to rise freely have been met, but as the warm air rises, the mountains provide an impenetrable obstacle to a northerly air flow. The rising air must therefore flow toward the equator. A surface flow from the equatorial latitudes of the Indian Ocean is established to replace the rising air over the northern portions of India. This warm air flow, known as the summer monsoon, originates over an extremely warm ocean and its water vapor content is at near saturation levels. The leading edge of the southwesterly flow is called the "burst" of monsoons. It brings the much needed rains to a region that has been precipitation free for almost six months. By late summer, the clouds and rain associated with the summer monsoon cool the region. The southwest monsoon flow weakens and the normal planetary wind field begins to reestablish itself over most of the region by late September. All of the conditions for the monsoon are reversed and high pressure, subsiding air, and northeast winds dominate the region from fall to spring when the cycle is renewed.

The Indian monsoon is the most significant macroscale wind reversal that occurs on earth. The West African monsoon along the Nigeria-Ghana coast occurs during the same period of the year (and for basically the same reasons) but is much more localized. Its effects are restricted to within 200 km of the coast.

Most of our scientific expertise has been directed at mapping and understanding the macroscale flows on earth and the other planets and moons with atmospheres. We may find examples of regional wind reversals, similar to but not exactly like the monsoons, on other planets as our data base increases. Similarly, it is nearly impossible to summarize the characteristics of the earth's global circulation in one chapter without selective omissions. Therefore, readers are urged to consult any of the basic meteorology textbooks if they wish a more comprehensive coverage of any of the topics associated with atmospheric circulations.

Condensates and Clouds

Condensation processes in the atmosphere are complex and highly variable. The factors which influence the actual phase change from water vapor to liquid droplets or crystals are vapor pressure and temperature. The two most common condensation nuclei are dust particles and sea salt crystals. The dust particles are advected from the ground into the atmosphere during bubble-breaking processes and are also carried aloft by the winds. It is generally safe to assume that sufficient condensation nuclei are always present in the earth's atmosphere.

Vapor pressure is a function of the water vapor content of the atmosphere. Water vapor is a gas and it exerts a pressure. A portion (≈2 mb) of the average surface barometric pressure

(1013.2 mb) is due to the partial pressure of water vapor. Saturation levels result when the number of water vapor molecules condensing is balanced by an equal number of water molecules evaporating into the atmosphere. Saturation levels are approached at relative humidities ≈100 percent, however, it is common for condensation to begin at relative humidities of <80 percent near the earth's surface. Condensation aloft may occur at supersaturation levels at relative humidities ≈101.5 percent.

Temperature relationships govern the atmosphere's capacity to hold water vapor. The capacity increases as the air temperature increases. Conversely, where air rises and cools, its capacity to hold water vapor decreases and the relative humidity increases. The air reaches saturation levels with continued cooling. The temperature at which saturation levels occur is the dew point temperature of the air mass. Condensation of cloud droplets occurs if the temperatures decrease any farther. These tiny water droplets may exist in the supercooled state at temperatures $\approx 40^{O}K$ below the freezing temperature of water. Ice crystals form as water vapor sublimates out of the atmosphere at below freezing temperatures.

The Growth of Cloud Droplets

The growth of cloud droplets is controlled by the solute effect and the curvature effect. All cloud droplets condense out onto condensation (hygroscopic) nuclei and dissolve some of the nuclei's chemical constituents. The solute effect results in a change in vapor pressure. Solution droplets have lower vapor pressures than pure water. The chemicals in solution interfere with evaporation from the cloud droplets. The interference may be physical, or it may be due to molecular chemical attractions. The concentration of the solution is equally important. Those cloud droplets which have dissolved more of the condensation nuclei will contain more concentrated solutions. These concentrated droplets will have lower vapor pressures and be less likely to reevaporate. The larger diameter droplets contain more water and are, therefore, less concentrated or more dilute solutions. They reevaporate more easily back into the atmosphere because of the solute effect.

The curvature effect is associated with the actual curvature of the outer skin of the cloud droplet. It takes more work to maintain a sharply curving small droplet than it does the more gentle curvature associated with the skin of the large droplets. The number of water molecules available to build water molecule structures is reduced on a sharply curved small droplet. Lesser molecular attraction translates into more rapid evaporation and a higher vapor pressure.

Growth of the droplet is determined by a balance between the solute and curvature effects. The solute effect diminishes as the droplet grows, i.e. the solution within the droplet becomes dilute. Growth lessens the radius of curvature, the vapor pressure, and the tendency to reevaporate. The droplet's vapor pressure increases with dilution and is balanced by a vapor pressure decrease with increased size. The net result is that cloud droplets grow to sizes ranging in diameter from 1-200 x 10^{-3} mm, generally averaging about 20 x 10^{-3}mm.

Ice crystal clouds form at temperatures below freezing. The shape of the ice crystals is determined by the temperature range at the moment of sublimation. In general, hexagonal crystals will form if temperatures are just below freezing. Temperatures approximately 10°-15°K below freezing produce more of the typical radiating, snowflake pattern. Temperatures <25°K below freezing produce ice prisms. All forms of crystalline water diffuse light prismatically. This characteristic is useful in determining whether ice crystals or water droplets dominate the upper cloud regions. The boundary between the blue sky and the edge of the cloud tops is the key to determining whether the upper cloud region contains water droplets or ice crystals. If the border is sharp, water droplets predominate. Ice crystals predominate if the border is diffuse or "fuzzy."

The diameter and mass of the typical cloud droplet is so small that it does not fall to earth. The turbulence of the atmosphere is sufficient to buoy it up. Since the major difference between cloud drops and raindrops (0.4 to 4 mm) is size, there are obviously additional processes which influence the transition from cloud droplets to raindrops. Otherwise, precipitation would be rare on earth.

<u>The Growth of Raindrops</u>

There are two processes that control the growth of cloud droplets into raindrops: coalescence and the Bergeron-Findeisen process. The coalescence process occurs inside clouds composed primarily of water droplets. The process proceeds in a sequential manner. Turbulence within the cloud moves the tiny cloud droplets and random collisions between droplets result. The collisions produce larger diameter drops. Certain drops will randomly grow large enough to begin falling toward the earth, assuming teardrop shapes as they fall. During their descent, the bottom side of the drops collide with more droplets and grow by "direct capture." They also grow by "lee capture," as small droplets are drawn into the falling drop's wake. These are also absorbed by the drop as it continues to grow. The process of "lee" or "wake" capture occurs for the same reasons as small automobiles tend to be drawn into and toward the rear of a large, passing truck. The truck creates air currents flowing along its sides and around in a circular path back toward the center-rear of the truck. The droplet creates an identical air flow as it falls toward earth. Coalescence explains the formation of raindrops from cloud droplets but it does not explain what occurs in clouds with ice crystals and supercooled water droplets.

The pioneering work on mid-latitude raindrop growth processes was conducted by Bergeron and Findeisen. The conditions of the clouds they studied are typical of those found in the middle and polar latitudes for many months each year. The research into the behavior of supercooled droplets and ice crystals showed that the vapor pressure for ice is less than the vapor pressure of the supercooled water droplets. The liquid water molecules may evaporate more easily from the water droplets than those water molecules bound into the ice structures. The ice crystals grow at the expense of the supercooled water droplets which reevaporate. The larger ice crystals fall toward the earth. If cloud temperatures at lower altitudes are above freezing, the crystals will melt and form raindrops. If atmospheric temperatures are below freezing, the crystals will reach ground level and accumulate.

It is common to find both coalescence and the Bergeron-Findeisen processes operating in mid-latitude and polar clouds. In the tropics, coalescence is the primary mechanism of raindrop formation due to the warmer atmospheric temperatures at the surface and mid-tropospheric levels.

Precipitation Production Processes

There are a wide variety of precipitation forms that occur in the earth's atmosphere but there are only three ways precipitation may be produced: orographically, convectionally, and cyclonically. Orographic uplift is a response to topographic effects where a regional air flow is forced to rise up over mountains or a mountain range. The forced ascent cools that air at the dry adiabatic rate. The dew point temperature is often reached and clouds form. If the uplift continues, more clouds and precipitation may result. The amount of moisture condensed out of the atmosphere by an orographic situation varies depending on the original water vapor content of the air and the amount of vertical ascent. The Western Ghats of western penninsular India are 2,000-3,000 m high. During the summer months, they condense out over 150-200 cm of rain from the southwest monsoon flow on their windward slopes.

Convectional precipitation results from cumuliform clouds produced by the intense heating of the earth's surface and the resulting convection that ensues. The typical convectional storm is the cumulonimbus or thunderstorm cloud. The characteristics of this particular cloud and the life cycle of the thunderstorm will be explained in later sections dealing with clouds and storms on earth.

Cyclonic storms are rotating energy release mechanisms. The precipitation they produce is secondary to their role in the earth's energy budget processes. The characteristics of cyclonic storms vary but the end result is the vertical ascent of air and the condensation of water vapor. The specific characteristics of cyclonic storms will also be discussed in the section dealing with storms.

Forms of Precipitation

A number of different precipitation forms result from water's ability to change phases on earth. Precipitation forms are either solid or liquid water forms. The most common precipitation forms are drizzle and rain. The only difference between them is the diameter of the drop (Table 1.4). Drizzle drop diameters are significantly smaller than raindrops. Freezing rain or a "glaze" results when raindrops impact on below freezing surfaces at ground level. The water drop splatters and freezes coating the trees, streets, etc. with a layer of ice.

Snowflakes range from 4-20 mm in diameter while snow pellets vary from 2-5 mm. Sleet begins as a raindrop. It passes through a layer of air with below freezing temperatures and freezes. Sleet diameters are generally <5 mm.

TABLE 1.4

Characteristic sizes and concentrations of atmospheric constituents near sea level. (After Miller and Thompson, Meteorology, 3rd ed., courtesy Charles E. Merrill Publishing Company, 1979, page 10.)

Type	Diameter (mm)		Concentration (no./cm^3)	
	Range of Sizes	Typical	Range	Typical
Gas molecules	$(2.8\text{-}6.5) \times 10^{-7}$	3.8×10^{-7}		2.5×10^{19}
Small ions	$(0.15\text{-}1) \times 10^{-5}$		$(1\text{-}7) \times 10^{2}$	
Large ions	$(1\text{-}20) \times 10^{-5}$		$(2\text{-}20) \times 10^{3}$	
Small (Aitken) condensation nuclei	$(0.1\text{-}4) \times 10^{-4}$		$10\text{-}10^{5}$	10^{3}
Large nuclei	$(4\text{-}20) \times 10^{-4}$		$1\text{-}10^{3}$	10^{2}
Giant nuclei	$(20\text{-}1000) \times 10^{-4}$		$10^{-4}\text{-}10$	1
Dry haze	$(1\text{-}100) \times 10^{-4}$		$10^{3}\text{-}10^{5}$	
Fog and cloud droplets	$(1\text{-}200) \times 10^{-3}$	20×10^{-3}	25-600	300
Drizzle	$(2\text{-}40) \times 10^{-2}$	30×10^{-2}	1-10	
Raindrops	0.4-4	1	$10^{-3}\text{-}1$	
Snow crystals	0.5-5	2	<10	
Snowflakes	4-20	10	$10^{-3}\text{-}1$	
Hail	5-75+	15	$10^{-6}\text{-}10^{-1}$	

Characteristic Ranges in Size of Some Other Small Particles in the Atmosphere (mm)

Viruses:	$(0.3\text{-}5) \times 10^{-5}$	Coal dust:	$(0.1\text{-}10) \times 10^{-2}$
Bacteria:	$(0.3\text{-}30) \times 10^{-3}$	Pollen:	0.01-0.1
Sea salt:	$(0.25\text{-}5) \times 10^{-4}$	Soil:	$(0.1\text{-}2) \times 10^{-3}$
Smoke:	$(0.1\text{-}10) \times 10^{-4}$	Silt:	$(0.2\text{-}2) \times 10^{-2}$
		Sand:	0.022-2

Hail or hailstones consist of semi-concentric layers of ice and are generally produced by the strong, convective overturning associated with cumulonimbus clouds. Water droplets are carried vertically into the colder, upper cloud zones and frozen. They fall and collide with more liquid droplets in the lower, warmer sections of the cloud. Strong updrafts return the water coated ice pellet to the upper cloud regions and the water freezes. The strength of the updrafts in the clouds determines the number of repetitions up and down through the cloud and the ultimate size of the hailstone. One of the authors (J.P.B.), vividly recalls being physically beaten by golfball-sized hailstones as he bicycled home from a friend's house during the onset of a particularly violent thunderstorm on June 9, 1953. A few minutes after he entered his home, it was destroyed by a massive tornado spawned by that thunderstorm (Fig. 1.41). Hailstones generally range from 5-75 mm in diameter but softball-sized hail has been reported from sites in Texas and Kansas.

Cloud Types on Earth

There are numerous cloud types observed on earth, however, almost all of them may be classified as a member of one of two cloud families: stratiform and cumuliform. Clouds are classified on the basis of their morphology and vertical development. Stratiform clouds are nearly horizontal clouds that build in sheet-like layers. Stratiform clouds generally cover the entire sky. They are produced when a flow of warm moist air is forced to rise up and over a colder denser air mass. A wedge-like layer of various stratiform cloud types may form where the rising air begins condensing out cloud droplets. The angle of ascent is very small so that the clouds produced may extend 1500 km out in advance of that point where the air was forced to leave the surface and rise over the colder air.

The individual cloud names are determined by their altitudinal zones. The zones are defined as the low, middle, and high cloud regions found at 0-2 km, 2-6 km, and <6 km, respectively. The upper cloud boundary is normally the altitude of the tropopause.

Stratus are low, stratiform clouds (Fig. 1.21). They obscure the sun and generally produce extended periods of steady, often heavy, precipitation. Stratus clouds may also be referred to as nimbostratus, if precipitation occurs. Stratus clouds are thicker than the other stratiform types because they are the first to be condensed out of the atmosphere. Since most of the earth's water vapor is in the lower 3 km of the atmosphere, large quantities of water vapor are condensed at the low cloud levels. Stratus clouds are thick enough to completely obscure the sun and are the type commonly observed on rainy days in the middle latitudes.

Altostratus clouds form at the middle cloud levels (Fig. 1.22). The air which rises to produce the stratus cloud continues to rise up over the colder, denser air. The bulk of the water vapor has already condensed out so the altostratus clouds are thinner, even though they are also sheet-like in their coverage of the sky. The sun's outline may be seen through altostratus clouds. Altostratus clouds may occasionally condense out some sporadic, light precipitation.

Cirrostratus clouds are the high cloud variety of stratiform clouds (Fig. 1.23). The rising air, which produced the stratus and altostratus clouds, continues upward condensing out a thin sheet of milky-white clouds. Almost all of the water vapor has now condensed out of the rising air. Sunlight easily shines through the cirrostratus clouds. Cirrostratus clouds consist of ice crystals as temperatures above 6 km are too cold for the formation of water droplet clouds. Often the ice crystals refract the sunlight (moonlight) and produce the familiar ring around the sun (moon).

The air flow continues upward condensing out what little moisture remains in the form of thin wispy filaments of ice crystal clouds known as cirrus clouds (Fig. 1.24). Cirrus clouds do not belong to either cloud family and may occur whenever small amounts of ice crystals condense out at high altitudes. They are a simple but unique cloud form.

Members of the cumuliform cloud family are identified by their vertical morphology. Their characteristics range from small, puffy, white clouds near the ground to massive, vertical forms towering from near ground level up to the tropopause. Cumuliform clouds are products of convection, i.e. rising columns of warm air and cold fronts. Convection is, in turn, dependent on the intensity of the solar heating of the earth's surface. The more intense the convection, the greater the vertical extent of the cumuliform cloud. Cold fronts also initiate strong, vertical motions in the atmosphere and may also trigger cumuliform cloud development.

The simplest form of cumiliform cloud is the cumulus humilis or "humble" cumulus (Fig. 1.25). Their appearance is often described as "puffy cottonball clouds." Their base is often 300-500 m above the ground and their total vertical extent is 200-300 m. They are produced by modest convection under stable atmospheric conditions and are typically referred to as "fair weather" clouds.

More intense convection may produce turret-like or castle-like clouds referred to as cumulus castillanus (Fig. 1.26). Their base often forms at levels similar to the cumulus humilis but their turrets may reach altitudes >1,000 m.

Towering cumuliform clouds are often produced several hours after noon on hot summer days when the air temperature is approaching its diurnal maximum value (Fig. 1.27). These towering clouds are termed cumulus congestus and have a "heaped" appearance. The base of these clouds may be several kilometers in diameter. The cumulus congestus clouds may begin forming at altitudes of 500-1,000 m and build up to 10,000-12,000 m. These clouds represent either intense convective activity or very strong vertical uplift. They often produce brief showers and indicate an atmospheric potential for thunderstorm development.

The largest cumuliform cloud is the cumulonimbus or thunderhead (Fig. 1.28). Cumulonimbi are often 10 km in diameter and reach altitudes of 15 km. Their tops often appear similar in shape to a blacksmith's anvil. The faster winds aloft shear the top of the cloud off and carry it downwind. When this happens, the top of the anvil is elongated and represents the approximate direction of the cloud's motion. Thunderheads appear as anvil-shaped or mushroom-shaped plumes that tower above the surrounding cloud masses in satellite photographs. High altitude plume clouds have also been observed on Jupiter. Their form also suggests processes associated with strong vertical motions.

The cumiliform clouds in the preceeding discussion may exist independently of one another, however, it is common to see all four types develop on a hot, summer day. Typically, the cumulus humilis and the cumulus castillanus types may develop within a few hours of each other. Similarly, one rarely observes a cumulonimbus without also observing numerous cumulus congestus clouds.

Fig. 1.21. Stratus clouds. Photographs by J. Barbato unless otherwise noted.)

Fig. 1.22. Altostratus clouds.

Fig. 1.23. Cirrostratus clouds.

Fig. 1.24. Cirrus clouds with cirrostratus near the horizon.

Fig. 1.25. Cumulus humilis or fair weather cumulus.

Fig. 1.26. Cumulus castillanus or "turret" clouds.

Fig. 1.27. Cumulus congestus clouds.

Fig. 1.28. Cumulonimbus cloud (thunderhead). This particular cloud was over 30 miles away when the picture was taken. It produced a tornado that struck the Lowell-Chelmsford, Mass. area in 1972.

Fig. 1.29. Stratocumulus clouds.

Fig. 1.30. Altocumulus clouds.

Cold frontal passages often produce cumulus congestus and cumulonimbus clouds. The specific types of clouds produced by an individual cold front are dependent on the temperature and moisture contrasts between the air in advance of the front and the air behind the front.

There are a number of intermediate cloud types identified by their combination of cumuliform-stratiform characteristics. Their morphology is puffy, similar to members of the cumuliform family yet they develop a limited, sheet-like appearance characteristic of stratiform clouds. The three intermediate cloud types are also classified according to their altitudinal zones. The low, middle, and high forms are termed stratocumulus, altocumulus, and cirrocumulus, respectively.

Stratocumulus clouds are often cigar-shaped and appear in echelon across the sky (Fig. 1.29). They form as cold, dry air passes over a warmer land surface. The air is warmed and becomes less stable. Limited vertical motions cause air to rise and condense out elongated clouds parallel to the prevailing wind direction. Stratocumulus clouds are typical during the fall months in the middle latitudes.

The second intermediate cloud type is the altocumulus cloud. It occurs at altitudes of 2-6 km and appears as an assemblage of white, puffy elements adjacent to one another with patches of blue sky visible through the assemblage (Fig. 1.30). These may appear anytime a thin layer of sheet-like clouds has condensed out in an atmosphere experiencing modest vertical motions sufficient to produce the puffiness associated with altocumulus clouds.

The third intermediate cloud type is the cirrocumulus cloud. It forms at altitudes >6 km. They are very thin clouds because there is so little moisture at these altitudes. They appear identical and form for the same reasons as the altocumulus types except that they are found at higher elevations. Their morphology is also identical but the diameters of the "puffs" which make up the cloud elements are much smaller. Cirrocumulus clouds are seen occasionally. They are not rare but occur much less frequently than the stratocumulus and altocumulus varieties.

The vast majority of clouds are found in the troposphere between sea level and 15 km. There are, on occasion, thin clouds observed near sunset, at altitudes of 20-30 km and 70-90 km. The clouds at 20-30 km in the stratosphere are nacreous clouds and become visible as the setting sun illuminates them. They often appear to have the full range of spectral colors, hence the name nacreous or "mother of pearl" clouds. The clouds at 70-90 km in the mesosphere are noctilucent clouds. Their composition is thought to be either ice crystals or fine cosmic dust that has accumulated in a layer near the upper portion of the mesospere. Scientists are still uncertain what causes both of these high-altitude varieties of clouds. Their effect on incoming solar radiation is negligible when compared to the clouds of the troposphere.

The tropospheric clouds introduced in the preceeding section represent the most common forms of earth clouds. There are many more variations, as one glance at a cloud atlas will show, however, the purpose of the cloud discussion here is to familiarize the reader with the more important varieties.

Surface Level Condensates

Fogs are stratus clouds "sitting" on the earth. Fog droplet diameters occupy the lower end of the diameter range for cloud droplets, i.e. 1-200 x 10^{-3} mm. They condense out of the atmosphere when air temperatures at the earth's surface reach the dew point temperature. There are four types of fogs on earth: radiation, advection, upslope, and frontal. Radiation fogs occur on nights with strong radiational cooling. The air above the ground is chilled to its dew point temperature. If radiational cooling continues, the lower layer ($\approx$10 m) may condense water vapor and produce a fog. If dew point temperatures are below freezing, frost will form. The actual fog thickness is dependent on the intensity of the radiational cooling and the air's humidity. Radiation fogs are commonly observed filling valleys in early fall. The cold air produced by radiational cooling drains into the valley bottoms and reaches its dew point. These are often called valley fogs.

Advection fogs occur when warmer air is physically transported over a colder surface. The air cools and a fog results. Normally, the cooler surface is a land surface, however, advection fogs are also produced when warmer air passes over and is chilled by cold ocean water.

Upslope fogs occur in regions where air flows up gradual

topographic slopes such as in the Great Plains region of the United States. The air cools as its elevation increases and a fog forms.

Frontal fogs form in association with warm fronts in middle latitudes. The heavy rain, particularly in the spring, humidifies the air but the surface is quite cold, often still frozen. The air at the warm front is warmer but the lower layer is chilled by the frozen ground forming a frontal fog.

The radiation fogs are most common over the temperate and polar land surfaces while advection fogs are more common to coasts with offshore winds and cold water environments. Frontal and upslope fogs occur less frequently. Condensates, whether in the form of clouds, fogs, dew, or frost, are commonplace on earth. We shall see that they also occur elsewhere in our solar system and that they are not always associated with water.

Condensates and Air Pollution

Condensates need some type of solid "surface" onto which they may condense. Cloud and fog droplets condense around such hygroscopic nuclei as dust particles and sea salt crystals. Dew and frost condensates develop on any surface at or near the ground, i.e. blades of grass, etc. The formation of clouds and fog are related to the concentration of atmospheric particulates. We are also aware of the role particulates play in the absorption of infrared energy. Nature and the planetary winds provide an ample supply of these nuclei to the atmosphere. Inadvertently, man began to assist nature through wood fires, coal burning, and industrial development. It is virtually impossible to assess man's total interference or impact on the atmospheric environment but it is certainly significant.

Smoke particulates from our factories and coal dust (fly ash), particularly from our electric generating facilities, introduce tons of condensation nuclei into the atmosphere. The problem associated with fly ash will become more severe as nations convert their power plants from petroleum to coal, wherever possible. Oxides of sulfur (SO_2 and SO_3) are released from coal or oil burning processes. The sulfur dioxide (SO_2) and sulfur trioxide (SO_3) react with water in cloud droplets and raindrops to form sulfurous and sulfuric acid.

$$SO_2 + H_2O \rightarrow H_2SO_3 \quad \text{(sulfurous acid)}$$
$$SO_2 + H_2O \rightarrow H_2SO_4 \quad \text{(sulfuric acid)}$$

The problems associated with acid rainfall have just begun to be observed. This problem already exists in most of the northeastern United States where species of fish are dying, unable to tolerate more acidic pH values of the water.

Similar deleterious effects have been observed with nitrogen, a normally unreactive gas in our lower atmosphere. Nitrogen oxides are produced at high temperatures similar to those occurring inside automobile engines and large furnaces or power plants. Three oxides of nitrogen, nitric oxide (NO), nitrous oxide (N_2O), and nitrogen dioxide (NO_2) are released to the atmosphere. Collectively, these three oxides are referred to as the NO_x compounds. They are chemically active and enter into other reactions. A typical reaction is the combination of nitrogen dioxide (NO_2) with water (H_2O) to produce nitric acid droplets.

$$NO_2 + H_2O \rightarrow HNO_3 \quad \text{(nitric acid)}$$

The nitric acid and sulfuric acid droplets irritate respiratory systems and chemically attack our homes and buildings. The NO_x compounds are even more insidious when their combined effects are noted. NO_x gases absorb incoming solar energy selectively in the blue wavelengths. The result is a yellow-brown gas cloud visible over most of the world's metropolitan centers during the morning hours. This brownish cloud is photochemical smog. Within the smog cloud, chemical reactions create other powerful oxidants which also affect the human body and our material possessions. It is impossible to list the numerous other pollutants or chemicals that our technological society pumps into the atmosphere. It is even more frightening to think of their potential effect(s) on the earth's energy budget and/or the planetary circulation.

A more detailed discussion of air pollution problems is presented elsewhere (see references at the end of this chapter). The introduction of these substances into the atmosphere is a matter for grave concern. In the context of this work, we simply do not have the space or time to adequately present this material and the potential danger it poses to the earth's energy budget relationships.

Storms on Earth

The need for meridional energy transfers on earth has already been established in a preceeding section of this chapter. Storms are born for precisely the same reason, i.e. the need to redistribute energy. We have already established that the earth's 23.5° tilt is responsible for the observed seasonal variations in the amount of solar energy received at the earth's surface. Since the greatest energy receipt is in the equatorial latitudes, air there is warmer and more buoyant (less dense). It tends to rise creating a belt of equatorial low pressure. In the subtropics, subsiding air produces a belt of high pressure. In the mid-latitudes, warm, moist air from the subtropics and cold, dry air from the polar regions collide periodically. The less dense, warm, moist air is forced to rise up over the colder, denser air. The rising air motion decreases the atmospheric pressure and creates the mid-latitude, low pressure belt. Finally, the nearly persistent radiational cooling of the polar land masses generates large quantities of dry, cold, dense air responsible for the polar belt of high pressure. The planetary wind field, previously discussed, flows in response to the pressure differences established by the differential heating of the earth's surface.

It has also been noted that the earth receives a surplus of energy in the tropics and experiences a deficit at the poles as more energy is radiated to space than is received from the sun. The winds flow in response to pressure differences induced by this thermal imbalance and transfer energy from surplus regions to deficit areas. There are several significant forms of rotating, cyclonic storms which transfer huge quantities of latent heat into the upper troposphere: (1) the tropical cyclone (the hurricane or typhoon) and (2) the extratropical cyclone (the mid-latitude low pressure system).

Cyclonic rotation and air flow in the northern hemisphere is in toward a center of low and up and out in a counter-clockwise spiral. All cyclones are massive heat engines born out of the need to redistribute energy. Their circulations are powered by latent heat release and they die when their energy supply is either interrupted or the regional energy imbalances have been adjusted closer to an equilibrium state.

There are significant differences in the factors which initiate the development of tropical and extratropical cyclones. Strong air mass contrasts in temperature and water vapor content prevail in the mid-latitudes. These contrasts are absent in the tropics where the air is classified as maritime tropical air. Fronts, created along the boundary zones of contrasting mid-latitude air masses, do not exist in the tropics. Barometric pressure differences common to mid-latitude sites are absent at tropical locations. The barometric pressure trace, except during cyclonic storms, is virtually the same from day to day. The most significant event of the day is the pressure change associated with the passage of the atmospheric tide. The weather is so repititious that one wonders about the processes that initiate "bad" weather.

There are a variety of ways precipitation is produced in tropical and subtropical locations. Orographic precipitation results when the moisture laden winds are forced to rise over high topography on the windward sides of islands or mountain ranges. Orographic uplift may even trigger thunderstorm development if the uplift is vigorous and the instability characteristics are favorable for the vertical development of cumulonimbi. Tropical land areas experience intense, solar heating and convectional thunderstorms are common, occurring on a daily basis in locations in the Amazon Basin.

The easterly wave is a unique weather pattern associated with the regions situated between the southern portion of the subtropical high pressure belt and the equator. Areas such as the Caribbean experience these waves which last for 3-4 days and cause periods of heavy tropical showers.

The development of an easterly wave is associated with the rate of subsiding air in the semi-permanent, oceanic, high pressure systems which lie centered at 25° north and south latitude. The air flow equatorward from the southern flanks of these high pressure systems is known as the Northeast Trade Winds in the northern hemisphere. The trade winds are constant from the northeast approximately 80 percent of the year. Wind speeds in the trade winds vary from 10 km hr^{-1} in the summer to 20 km hr^{-1} during the winter months. During the late summer and fall, the trade wind speed varies as the intensity of the subsidence in the high pressure system varies. Periodically (during the late summer and early fall), the subsidence within the high intensifies and air moving faster than 10 km hr^{-1} emerges from the high and flows equatorward. This faster moving air overtakes the slower moving trade wind air and collides with it. The air in the collision zone is forced to rise creating a zone of ascending air or a bump along the axis of the wave (Fig. 1.31). This zone delimits the actual location of the wave. The ascending air rises, cools, and produces stratocumulus, cumulus congestus, and

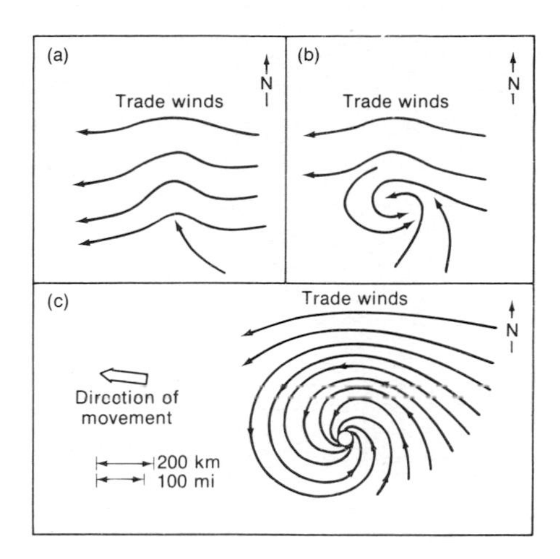

Fig. 1.31. An easterly wave (a & b) and the development of a tropical cyclone. (Courtesy Miller & Thompson, Meteorology, 3rd Ed., Charles E. Merrill Publ. Co., 1979, p. 172.)

cumulonimbus clouds and the three to four days of heavy tropical rains associated with the easterly wave. If upper tropospheric flows are moving away from the area above the surface easterly wave, the rising air will be encouraged to continue rising to fill the "void" created by the upper air movements. These circumstances mark those occasions when the upper troposphere supports the motion of the ascending air from the surface along the easterly wave. These conditions are referred to as "upper level support." Upper level support increases the potential for tropical storm development and periodically, the initiation of cyclonic rotation in the atmosphere (Fig. 1.31). Intensification of the now cyclonically rotating vortex is dependent on the release of latent heat energy. As it intensifies, it is called a tropical storm. It is termed a tropical cyclone (hurricane/typhoon) when winds >125 km hr^{-1} occur.

Tropical Cyclones

Tropical cyclones develop over the tropical oceans 5°-8° north and south of the equator. They cannot originate any closer to the equator because the Coriolis effect is virtually non-existent at these latitudes. Rotational motions are impossible in the absence of the Coriolis effect. Tropical cyclones develop from easterly waves and require the existence of upper level support.

The tropical cyclone draws hot, moisture-laden air in toward its center (Fig. 1.32). It forces the air to rise and releases vast

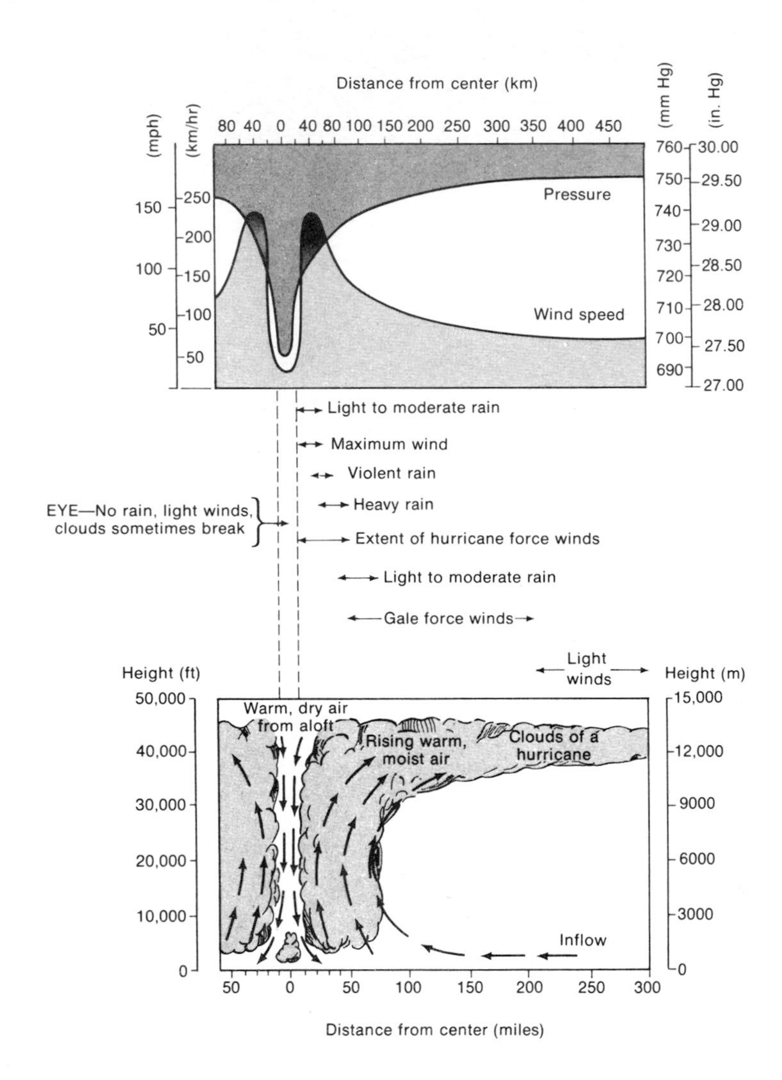

Fig. 1.32. Vertical cross section of a tropical cyclone and associated wind pressure and rain patterns. (Courtesy Miller & Thompson, Meteorology, 3rd Ed., Charles E. Merrill Publ. Co., 1979, p. 172.)

quantities of latent heat energy. The rising air is warmer at each successive level than air around the hurricane so it continues to rise explosively forming the dense lower clouds and upper cloud shield depicted in Fig. 1.32. The air near the top of the storm is virtually moisture free. All its water vapor has condensed out at lower levels. The cyclone's vertical circulation is so intense that this "dry" air is injected into the lower stratosphere.

Tropical cyclones are usually 300-350 km in diameter, however, the associated winds, clouds, and precipitation effects may extend great distances from its center. Wind velocities are typically >180-200 km hr^{-1}. The intensity of a tropical cyclone as well as its diameter is, in part, determined by the length of time it remains over the warm, tropical oceans. The tropical cyclones (typhoons) which strike the Philippine Islands and Japan are much larger than their Atlantic Ocean counterparts because the equatorial Pacific is geographically much larger. The storms grow larger and more intense before they finally move inland in the western Pacific region (Fig. 1.33).

The central portion of the tropical cyclone is the eye. The eye is generally <40 km in diameter. The eye is a relatively cloud-free region due to subsiding air from the stratosphere. The air subsides in response to the injection of air into the stratosphere from the cloud regions of the storm. Subsiding air warms at the dry adiabatic lapse rate. Clouds cannot form unless condensation occurs. Since warming air holds increasing amounts of water vapor, the eye is a distinct, cloud-free region of the tropical cyclone. There is also a distinct zonation of meteorological phenomena within a hurricane. The strongest winds, heaviest rains, and greatest release of latent heat energy all occur within 40-50 km of the eye. The clouds, increased wind speeds, and the large regional drop in atmospheric pressure may extend up to 500 km from the storm's eye.

The prevailing Northeast Trade Winds steer the tropical cyclones westward in the northern hemisphere. In many instances, the storms will pass inland, however, they often curve northward and are entrained into and steered by the westerly wind belt. It is generally expected that a tropical cyclone spawned in the Atlantic Ocean will pass across the Caribbean Sea and either move inland over Mexico or the Texas Gulf Coast or recurve northward and track along the Atlantic seaboard.

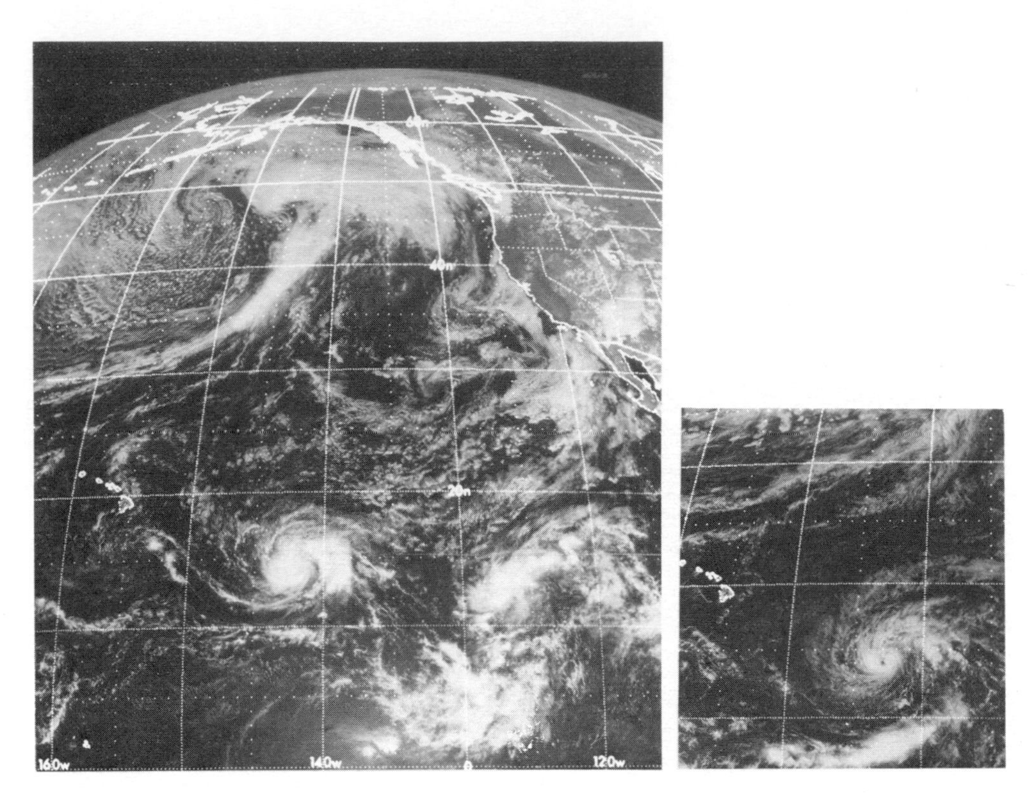

Fig. 1.33. Tropical cyclone in the Pacific Ocean, 23 Sept. 1976. Two days later, storm had a well developed eye (inset). (Note the mature extratropical cyclone south of Alaska.) (Courtesy NOAA.)

A tropical cyclone dissipates when it moves inland. Deprived of the warm, moist, ocean air which is its energy supply, it usually loses its well-defined physical structure within 12 hours. Twenty-four hours after passing inland, the remnants of the cyclone are usually nothing more than a large geographic area of clouds, heavy rain, and decreasing winds.

The structure of tropical cyclones is idential in both hemispheres, only the rotational motion differs (i.e. in the southern hemisphere cyclones rotate clockwise, in toward the center and up). We have already noted that the typhoons of the Pacific are generally larger due to the size of the Pacific Ocean and its temperature regime. The great expanse of water permits the storms to grow larger before striking land. Also, the North Pacific Ocean does not exchange large quantities of water with the Arctic Ocean. It is therefore 1°-2°K warmer than the Atlantic Ocean. The tropical Pacific Ocean provides a prodigous source of latent heat energy for the typhoons to draw upon.

Hurricanes in the Atlantic Ocean and typhoons in the Pacific do not have an exact counterpart in the Indian Ocean, however, small, intense, cyclonic storms known as Bay of Bengal Cyclones and Arabian Sea Cyclones do exist. These storms usually develop during the spring and fall (i.e. the transition seasons between the monsoon flows) and are between 100-160 km in diameter. They have hurricane force winds and are often accompanied by tremendous storm surges as they move inland. Many episodes of human misery and suffering are associated with these compact but devastating storms. In the Ganges River Delta of India and Bangla Desh, 300,000 people died from the wall of water which surged inland accompanying a Bay of Bengal cyclone in 1970. Another 200,000 died from cholera spread via the contaminated water supplies. Bay of Bengal and Arabian Sea Cyclones are the North Indian Ocean's counterpart to the hurricanes and typhoons of the Atlantic and Pacific and the energy exchange roles that they play in relation to the energy budget.

Tropical cyclones in any of the earth's oceans are interesting and dynamic rotational vortices. While they exhibit a distinct range of meteorological characteristics, these are secondary in nature. The tropical cyclone's primary role is the transfer of surplus energy from the earth's oceans and tropical atmosphere to the upper troposphere and lower stratosphere. This transfer of latent heat is a significant component of the earth's energy budget. In chapter five, we shall see that hurricane-like storms may also exist on Jupiter.

The Extratropical Cyclone

Seasonal variations in solar heating result in larger thermal contrasts in the middle latitudes. Diurnal and regional pressure differences result in higher average wind velocities. Radiational cooling over the northern continents during the fall, winter, and spring months creates huge masses of continental polar and arctic air. These cold air masses become so large that eventually, like an amoeba dividing, a portion of the cold, dry air is exported equatorward as a cold core high pressure system. These cold air masses move southward across the Great Plains and Mississippi Valley and then eastward as they are "caught" and steered by the prevailing westerly winds. These cold core highs provide a strong air mass contrast to the warmer air to the south. This sets the stage for air mass collisions, frontogenesis, i.e. the formation of fronts, and cyclogenesis, i.e. cyclone development.

The subtropical high pressure systems located over the world's oceans at latitudes 25°N and 25°S are referred to as warm core highs. They experience an increased rate of subsidence (sinking air) as the increased thermal gradients and pressure gradients during the spring, fall, and winter months accelerate atmospheric motions. The surface level response is an increase in wind speeds which emanate from all sides of the highs in the northern hemisphere. In the northern hemisphere, the air flow from the western flank of the oceanic highs transports warm, moist air poleward. Warm, moist, maritime tropical air from the Caribbean and southern North Atlantic is advected toward the eastern half of the United States. The poleward flow of warm, moist air eventually collides with the equatorward flow of cold, dry air along a broad zone or region of cyclogensis associated with the lcoation of the polar front.

Whenever air masses of differing temperature and moisture characteristics collide, fronts are formed. A front is the zone separating the contrasting air masses. Air mass contrasts are based on temperature, humidity, and pressure differences. There are four types of fronts associated with mid-latitude weather phenomena: (1) stationary, (2) cold, (3) warm, and (4) occluded. The stationary front (Fig. 1.34) is an areas of mixed clouds and showers between a cold, dry air mass and a warm, moist air mass. If the upper level conditions associated with the upper level westerlies support the rising air along the front, a cyclone will begin to form. Generally, the weather along the stationary front is poorly organized and varies with the intensity of the air mass contrasts.

The cold front (Fig. 1.35) is the leading edge of a large, cold core, high pressure system. The actual frontal zone is generally

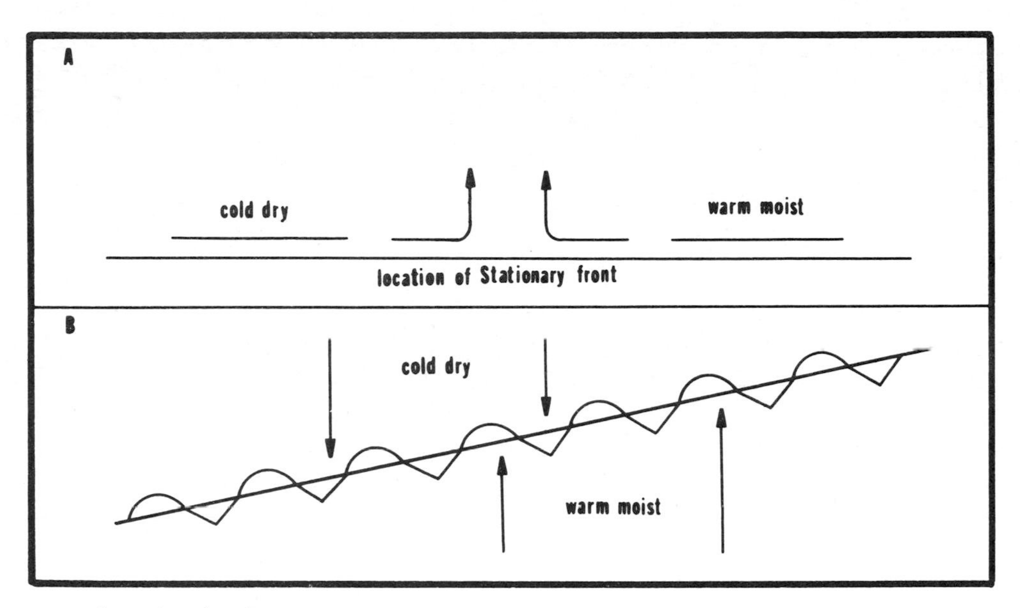

Fig. 1.34. Cross section and symbol for a stationary front.

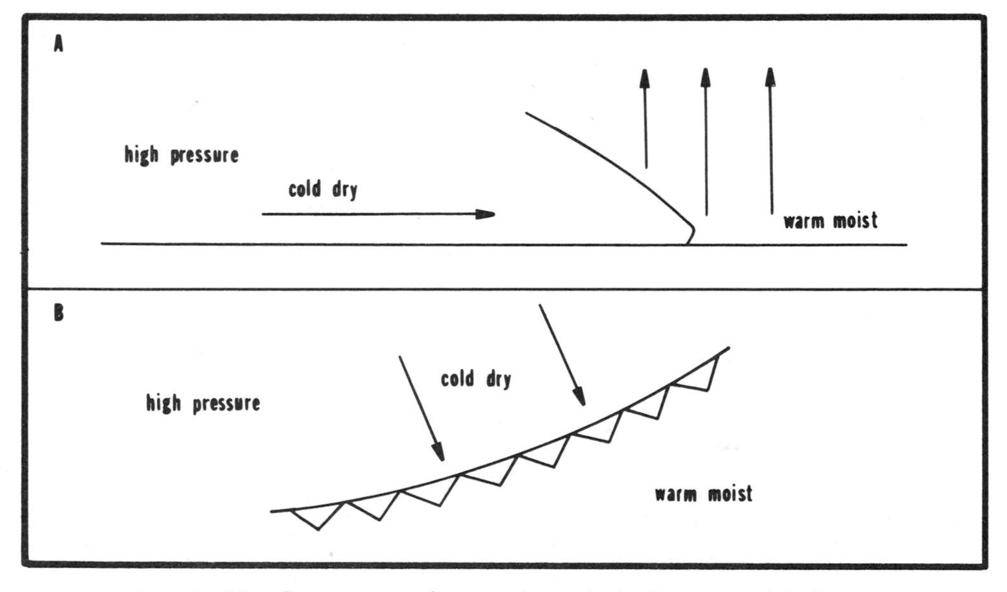

Fig. 1.35. Cross section and symbol for a cold front.

less than 80 km wide at ground level while the high pressure system of which it is a part may be larger than 1,000 km in diameter. It is impossible to identify the air temperatures that occur in a cold core high. They change seasonally but are always less than the temperature of the maritime tropical air masses. Similarly, the moisture levels vary seasonally. The lowest humidities in cold core high pressure systems occur during the late winter months in association with the coldest air. The cold air associated with the cold front usually underruns warm, moist air forcing it to rise. This vertical motion produces cumuliform clouds. Cumulonimbi (thunderstorms) and heavy showers may occur during the warmer months. It is common to see a line of thunderstorms develop along and just in front of the cold front. These "squall lines" are particularly significant due to the intensities of the cumulonimbi and the high tornado probability. Despite the severity of weather phenomena that may accompany a cold front, it generally passes across a site in one to three hours. The passage of the cold front is typically followed by rapidly clearing skies, northwest winds, and rising barometric pressure in the eastern United States.

A warm front forms when a warm, moist air mass flows poleward and collides with a cold, dry air mass. Since the cold air is denser, the warm, moist air is forced to leave the surface and rise up over the colder air (Fig. 1.36). The warm, moist air rises and cools. Its water vapor condenses out and forms stratus, altostratus, and cirrostratus clouds. Light precipitation may be produced by the lower altostratus and heavy, long-duration precipitation from the stratus clouds. The geographic extent of the warm front may range from 800-1,500 km. The duration of the precipitation may be from twelve to thirty-six hours depending on how fast the front progresses northward. The front is mapped where it intersects the ground. Its forward progress averages 10-15 km hr^{-1}.

An occluded front occurs when the cold front catches up with and underruns the warm front (Fig. 1.37). The remaining warm, moist air is uplifted and the moisture wrung out. With the termination of its energy supply of warm, moist air (latent heat), the clouds slowly dissipate as the front disintegrates.

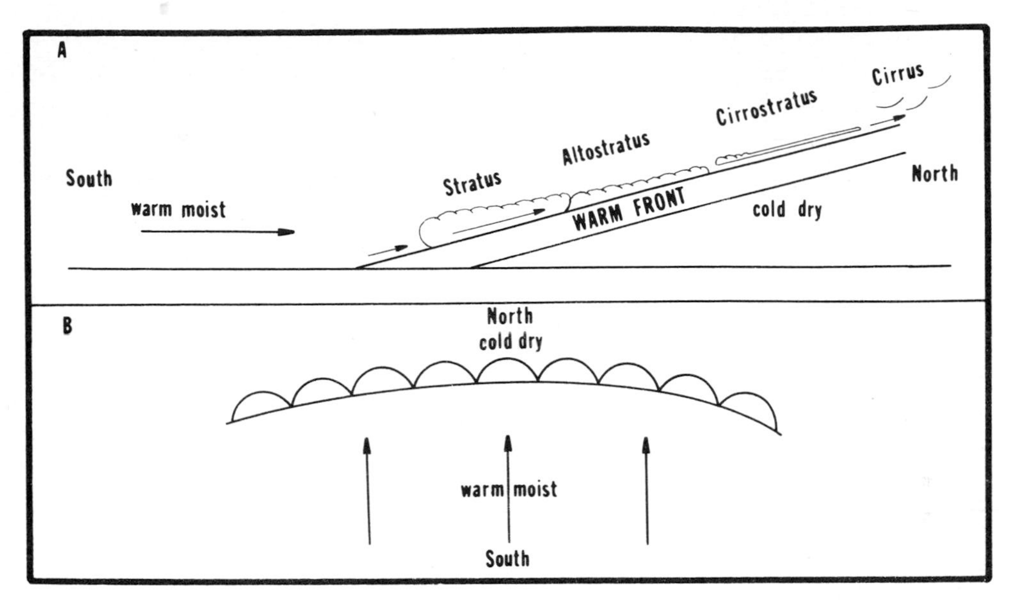

Fig. 1.36. Cross section and symbol for a warm front.

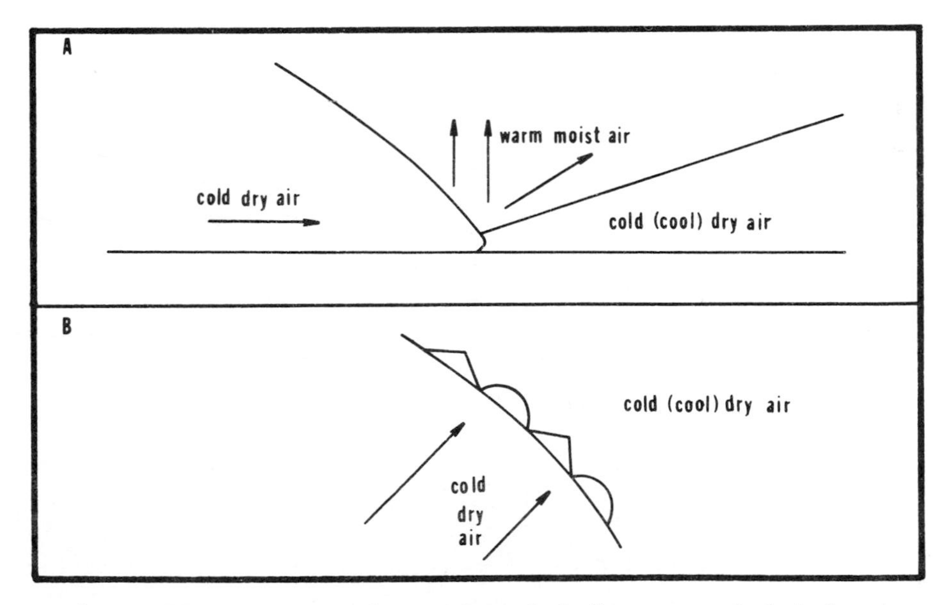

Fig. 1.37. Cross section and symbol for an occluded front.

The Life Cycle of an Extratropical Cyclone

The life cycle of a mid-latitude cyclone is depicted in Fig. 1.38, A-D. Stage A represents the collision of contrasting air masses to

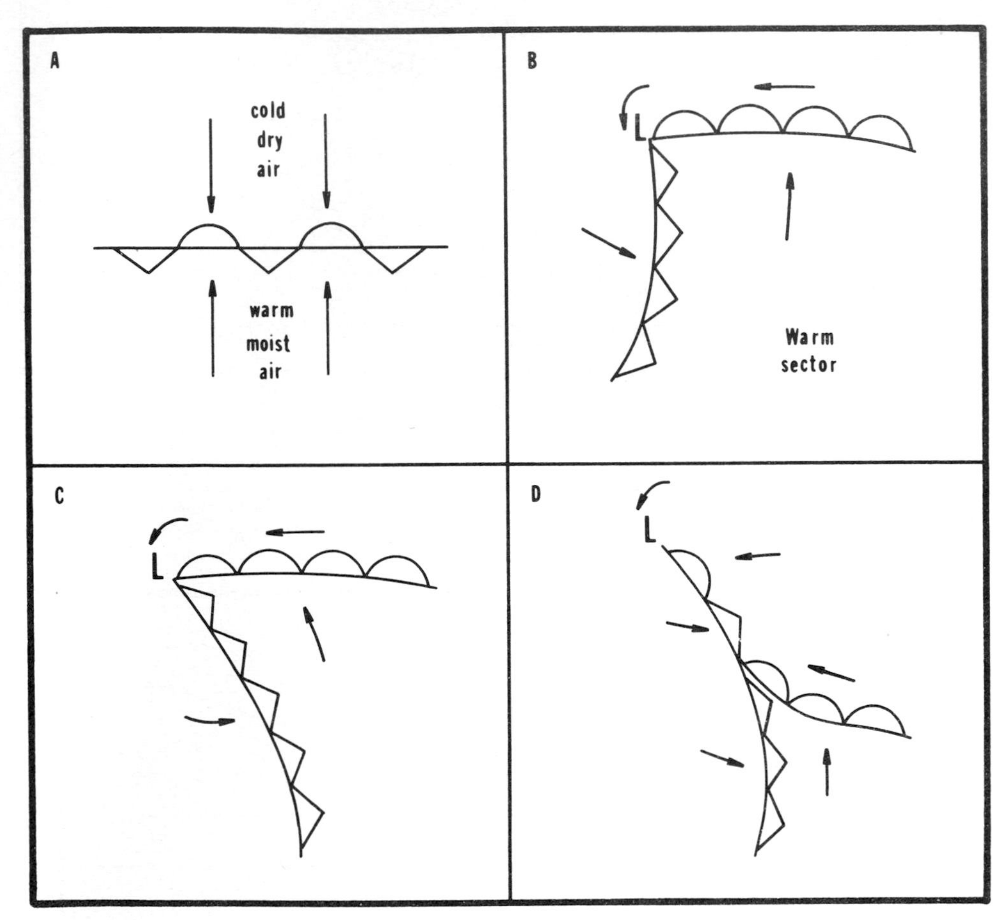

Fig. 1.38. The life cycle of a mid-latitude cyclone.

form a stationary front. The continental polar air to the north eventually begins advancing southward forming a cold front while the warm maritime tropical air advances northward forming a warm front (Stage B). At the same time, these two flows initiate cyclonic (counterclockwise) rotation in the atmosphere giving birth to a mid-latitude, low pressure system. This will only occur if upper level support is present aloft. Stage C is the mature stage. The storm is fully developed at this stage (Fig. 1.39). The storm's

Fig. 1.39. Satellite photograph of a mid-latitude cyclone located east of Newfoundland. The cold front trends NE↔SW. The stratiform cloud shield trends E↔W. See Fig. 1.33. (Courtesy NOAA.)

diameter may be >1,500 km with winds ranging from 20 km hr^{-1} to 125 km hr^{-1}, depending on the storm's intensity. The orientation of the fronts changes as the denser air at and behind the cold front moves faster than the warm front. The cold front begins to catch up with the warm front during Stage C. Stage D represents the death or occlusion of the storm. The cold front has overtaken and underrun the warm front, closing the warm sector or energy supply of the storm. When the cold front terminates this flow, the storm weakens and dissipates.

The primary role of these storms is the transfer of latent heat. Rain and/or snow are produced in the stratus clouds and vast quantities of latent heat are released at the moment of condensation as real heat. This release of heat acts to moderate the upper atmospheric temperature profile completing the process of latent heat transfer discussed previously in the section dealing with the earth's global energy budget.

Hurricanes and mid-latitude low pressure systems are the two most significant mechanisms whereby meridional energy exchanges occur. Substantial quantities of heat energy are transported poleward in the flows of warm ocean currents such as the Gulf Stream or Japan Current but these are secondary to the atmosphere's role in latent heat transfer processes.

Thunderstorms

Thunderstorms are very localized severe weather clouds often 15-18 km in diameter. They are initiated by (1) convection and (2) cold fronts. Convective thunderstorms are created as warm, moist air rises vigorously. These conditions are generally met on typical summer days between 3-6 P.M. local time. Cold front thunderstorms are triggered as the cold or cool, dry, polar air underruns warm, moist air forcing it to rise. The rising column of air remains warmer than the surrounding air at each successive level due to latent heat release. The vertical motions are often sufficiently large enough to form a cumulonimbus. The thunderhead generally flattens out, taking on the traditional "anvil" shape, as the rising air reaches the tropopause and spreads horizontally. The cold front thunderstorm may occur any hour of the day or night. It is totally dependent upon the passage of a cold front through a region.

A well developed cold front may spawn many thunderstorms along the axis of the cold front. This is termed a squall line. It is not uncommon for the squall line to outrun the cold front that created it and, in some cases, permit the development of a second squall line. Whether it be an individual thunderstorm or a series of squall line thunderstorms, the process is the same. Warm air is either forced to rise or rises due to convection, condenses, and releases latent heat.

The three stages of the life cycle of a thunderstorm are presented in Fig. 1.40. The young stage (a) consists of updrafts. As the thunderstorm matures, many of the falling raindrops reevaporate, cooling the air. The cool air becomes denser and sinks down and out the leading edge of the thunderstorm cloud. This process not only produces the strong downdrafts in the front portion of the mature stage cloud (b) but also is responsible for the squall-like conditions and the colder blast of air just prior to the onset of a thunderstorm. Typically, the air temperature decreases 5°-10°K as wind velocities increase. The mature stage represents the most

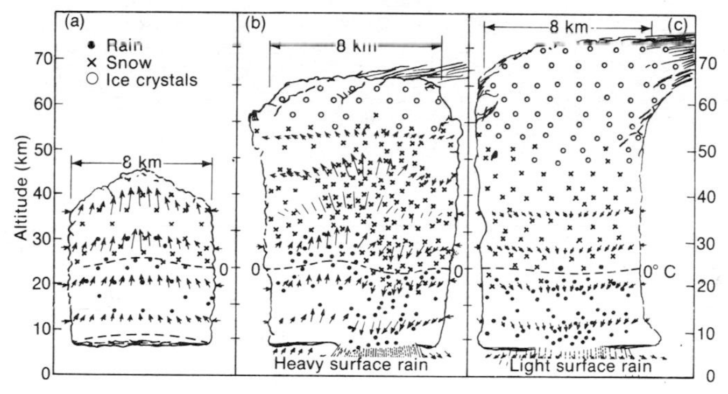

Fig. 1.40. The life cycle of a thunderstorm: birth (a); mature stage (b), and dissipating stage (c). (Courtesy Miller & Thompson, Meteorology, 3rd Ed., Charles E. Merrill Publ. Co., 1979, p. 146.)

violent stage of a thunderstorm where precipitation amounts are greatest, electrical activity is at its maximum, and wind velocities are very high. The strong updrafts within the cloud make hail formation possible while the cloud itself releases enormous quantities of latent heat energy to the upper troposphere. Finally in the dissipating state (c), the general air movement in the cloud is down as the thunderhead weakens and dissipates.

The cold blast of air in front of the thunderstorm is produced as denser air, cooled by the reevaporation of falling raindrops, sinks toward the earth's surface. It produces the decrease in air temperature and gusty winds mentioned earlier. It also helps perpetuate the storm as the blast of cooler air forces more warm, moist air ahead of the storm to rise and condense out moisture and release more latent heat energy. Convectional thunderstorms normally begin at or close to the time of maximum air temperature, i.e. 2-4 P.M. local time. They may continue until well after sunset. Finally, radiational cooling increases the stability of the air ahead of the storm by decreasing its temperature. The storm is unable to force more air to rise and dissipates.

Thunderstorms along the east coast of the United States often proceed until they reach the coast. The air over the water is cooler and hence more stable. The thunderstorms cannot force it to rise and the storm dissipates within a few kilometers of the shore.

There are 1,800-2,000 thunderstorms in existence on the earth at any moment. It is easy to identify their plume or anvil-shaped tops from satellite photographs. It is clearly evident that the latent heat released by thunderstorm activity is another important part of the global energy budget.

Thunderstorms also play an important role in the electrical balance of the earth. The earth is a large capacitor with the atmosphere charged positively (+) and the ground negatively (-). The atmosphere would normally lose its charge in a matter of a few hours without thunderstorms. Thunderstorms recharge the atmosphere. The actual charge distribution in a thunderstorm is segregated so that the bottom of the cumulonimbus has many surplus negative charges and the upper portion of the cloud is positively charged. The reasons for this distribution are still under investigation. It would appear that updrafts are stripping positive charges from water droplets in the cloud and carrying them aloft leaving the negative charges in the bottom half of the cloud. The negative charges in the lower cloud affect the charge distribution at the earth's surface. Normally, negative charges abound at ground level. The presence of a large, negatively charged cloud mass repels the negative ground charges, leaving the earth's surface beneath the cloud positively charged and setting the stage for a lightning discharge.

The electrical potential prior to a lightning bolt may reach 3,000 volts cm^{-1}. The differences between either end of the bolt may exceed hundreds of millions of volts.

The actual lightning bolt represents a series of successive steps beginning with a pilot leader moving down toward the ground from the cloud base. The air column is ionized and step leaders move farther down the ionized column. Streamers from the ground move up to complete the formation of an ionized air column. The negative charges flow downward in the main lightning stroke. The main stroke normally reduces the electrical gradient. Often, additional dart leaders follow the main stroke to complete the reduction in electrical potential. All of these steps occur in 2 x 10^{-4} seconds. The movement of negative charges from the cloud to the ground leaves the atmosphere positively charged. Thus, thunderstorms maintain positive charges in our atmosphere and preserve the electrical balance of our planet.

Tornadoes

Violent thunderstorms often spawn tornadoes. They are usually located near the rear portion of the cumulonimbus where vigorous updrafts initiate rotational motion. Many tornadoes remain up in the thunderstorm clouds but some descend to the earth's surface causing death and destruction.

Tornadoes are intense cyclonic vortices with extremely low pressures inside their funnel (Fig. 1.41). The greatest cause of damage is the drop in atmospheric pressure within the funnel. Houses where doors and windows were closed trapping air at higher atmospheric pressure, "explode" outward as the high pressure, inside air flows toward the intense, low pressure outside the house. Tornadoes are normally 100 m in diameter and contain winds estimated at 300-500 km hr^{-1}. They are commonly spawned by squall line thunderstorms.

The actual conditions necessary for the growth and development of severe thunderstorms and tornadoes would occupy a full volume. The reader is urged to consult any basic meteorology text for additional information.

Sandstorms

Weather satellites now provide us with detailed cloud photographs of hurricanes, mid-latitude low pressure systems, and the anvil tops of the thunderheads. They also allow man to view sandstorms or dust storms. Sandstorms and dust storms are not typical weather phenomena on earth but they are occasional meteorological events. During the Great Saharan Drought of 1970-1972, the wind carried terrestrial materials high into the atmosphere. Satellite photographs showed a cloud of dust extending from the western Sahara Desert, 4,800 km westward over the Atlantic Ocean to the Barbadoes

Fig. 1.41. Photograph of the tornado that struck Worcester, Mass. on June 9, 1953 killing over 200 people and causing millions of dollars in damage. (Courtesy of the Worcester Telegram and Gazette, Worcester, Mass.)

Islands in the Lesser Antilles. Historically, the "Dust Bowl" era of the 1930's in the United States was a period of severe dust storms caused by drought in combination with poor farming practices. While meteorological phenomena on earth, dust storms are typical on Mars and will be discussed in more detail in chapter four.

Summary

The earth's atmosphere is a dynamic gaseous envelope whose primary motions are initiated by pressure differences created by the sun's differential heating of its planetary surface in response to its 23.5^o axial tilt. The range of storms on earth are a response to the need to balance or "attempt to balance" the pattern of energy distribution on earth. All of the storms, except the sandstorm or dust storm, build water clouds and produce precipitation. Latent heat is geven off to the atmosphere at the moment of condensation. This completes an energy transfer cycle which began as evaporation of water over a tropical or subtropical ocean and culminated in the release of heat energy in the upper atmosphere over the mid-latitude regions.

We shall compare the processes and phenomena which occur on earth to those occurring in the atmospheres of other planets and moons. By comparing and contrasting with those things we have either experienced or observed on earth, our study of the dynamics of other planetary atmospheres should be easier and more meaningful. Finally, as scientists and students of science, we often forget to exercise one of the most precious gifts we have as human beings: aesthetic appreciation. Take the time as you read to appreciate the beauty evidenced by the nature of our solar system. As our study proceeds to the various planets and the dynamic processes of their planetary atmospheres unfold, remember the words of Sir Francis Bacon:

"In nature things move violently to their place,
and calmly in their place."
Bacon

References

The references for the earth's atmosphere are too numerous to list. The reader is encouraged to consult any of the basic meteorology and air pollution texts for additional information.

Butcher, S.S. and Charlson, R.J., An Introduction to Air Chemistry, Academic Press, New York, 1972.

Cloud, P.E., Atmospheric and hydrospheric evolution of the primative earth, Science, 160, 729 ff., 1968.

Eagleman, J.R., Meteorology: The Atmosphere in Action, Van Nostrand, New York, 1979.

Heicklen, J., Atmospheric Chemistry, Academic Press, New York, 1976.

Kerr, R.A., Origin of life: new ingredients suggested, Science, 210, 42 ff., 1980.

Lovelock, J.E. and Lodge, J.P., Oxygen in the contemporary atmosphere, Atmospheric Environment, 6, 575 ff., 1972.

Lovelock, J.E. and Margulis, L., Atmospheric homeostasis, by and for the biosphere: the Gaia hypothesis, Tellus, 26, 1 ff., 1974.

Margulis, L. and Lovelock, J.E., Biological modulation of the earth's atmosphere, Icarus, 21, 1 ff., 1974.

Miller, A. and Thompson, J., Elements of Meteorology, 3rd ed., Merrill, Columbus, Ohio, 1979.

Navarra, J.G., Atmosphere, Weather and Climate: An Introduction to Meteorology, Saunders, Philadelphia, 1979.

Watson, A. et al., Methanogenesis, fires and the regulation of atmospheric oxygen, Biosystems, 10, 293 ff., 1978.

CHAPTER II

MERCURY

Descriptive Statistics:

Distance from the Sun	58 million km 46 million km at perihelion 69 million km at aphelion
Diameter	4,830 km (3/8 earth's diameter)
Mass	1/20 of earth's mass
Mean Density	5.0 g cm^{-3}
Period of Rotation	58.6 days

Introduction

If the definition of a planetary atmosphere is a gaseous envelope which surrounds the planet, then Mercury has an atmosphere. Scientists may debate the nature of Mercury's atmosphere for decades because its atmosphere is extremely transitory. This transitory nature is due to two factors. First, Mercury's gravitational field is very weak. The escape velocity for gaseous molecules is only 4.0 km sec^{-1}. This value is low when compared to the 11 km sec^{-1} escape velocity for gaseous molecules in the earth's upper atmosphere. Secondly, that percentage of the planet's atmosphere which is attributable to the solar wind has not been determined. Certainly, the low gravity conditions on Mercury preclude anything but an extremely rarified atmosphere.

Atmospheric Composition

Optical studies of Mercury's halo have determined that, while extremely thin, Mercury's atmosphere extends upward 600 km from the surface. The primary gaseous compoment, based on the halo's brightness is hydrogen. The presence of this hydrogen was definately confirmed when the broad lines of atmonic hydrogen were identified spectroscopically from earth. The primary difficulty in studying the composition of Mercury's atmosphere arises from Mercury's position in the solar system. Mercury is never more than 28° from the sun. Given the earth's rotational speed of 15° per hour,

Mercury can rise at a maximum, almost two hours prior to earth's sunrise or set two hours after earth's sunset. Hence, telescopic observations of Mercury are constantly hindered by the brilliance of the nearby sun or made impossible because Mercury is positioned on the opposite side of the sun relative to the earth.

Results from Mariner 10's spectrometer have revealed the presence of hydrogen, helium, neon, argon, oxygen, xenon, and carbon in Mercury's atmosphere. The presence of atomic hydrogen was not surprising as Mercury's weak magnetic field freely allows proton fluxes from the sun to penetrate its atmosphere. Any atomic hydrogen lost by Mercury's atmosphere is continuously replenished by the flow from the sun. Mariner 10's ultraviolet spectrometer also detected neutral helium. These helium atoms have been identified as transient atoms from the solar wind. Their residence time, or the time each helium atom remains trapped in Mercury's atmosphere, is $\approx$200 days. Helium atoms are constantly being added to Mercury's atmosphere as others are being swept away by the force of the solar wind. The presence of argon, neon, and xenon had been suspected because of the weak luminosity detected from Mercury's dark side. Mariner 10 confirmed their presence.

Magnetic Properties and Magnetosphere

The Mariner 10 spacecraft had three separate encounters with Mercury between March, 1974 and March, 1975. The probe passed to within 705, 50,000, and 327 km of Mercury's surface during encounters 1-3, respectively. The spacecraft almost collided with Mercury during the third encounter. The first and third encounters yielded information that proved that Mercury did have a planetary magnetic field. The spacecraft did not pass close enough to the planet to observe the magnetic field's properties during the second encounter.

The presence of a planetary magnetic field had been suspected for some time. Mercury's mean density is approximately 5.0 g cm^{-3}, very dense for a planet only 4,830 km in diameter (3/8 earth's diameter). Density values so similar to the earth's mean density (5.5 g cm^{-3}) were interpreted to represent the presence of a large iron-nickel core, encompassing as much as 2/3 of the inner radius of the planet. Since the 2/3 ratio is greater than the ratio of the radius occupied by the earth's iron and nickel core, variations from a simple dipole magnetic field were anticipated for Mercury.

A model of Mercury's magnetosphere was contructed based on data provided by the first and third Mariner 10 encounters. The model (Fig. 2.1) clearly depicts a magnetosphere whose shape is remarkably

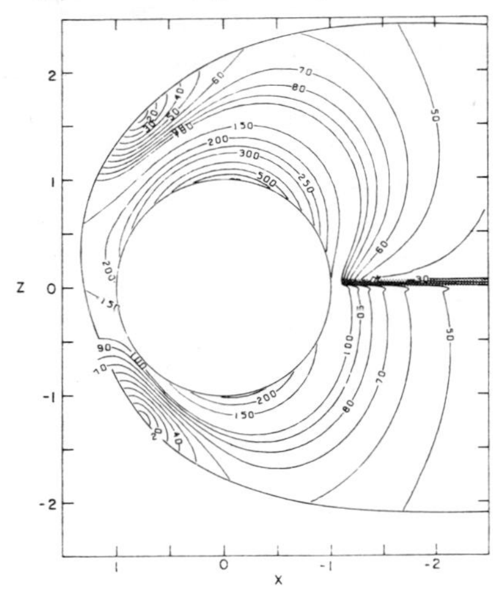

Fig. 2.1. A model of the magnetic field isointensity contours (gammas). (Courtesy Y.C. Whang, J. Geophys. Res., v 82, p. 1029, 1977, copyrighted by the American Geophysical Union.)

similar to the earth's magnetosphere. The compression of the field lines by the solar wind on the sun side is even more dramatic because of Mercury's proximity to the sun. Similarly, field lines on the opposite side depict an elongation of the magnetic field, again as a response to the interaction between the solar wind and the magnetic field. Thus, a compression or bow shock marks the edge of the magnetosheath on the sun side while a magnetic tail, similar in shape to the earth's magnetotail, exists on the opposite side.

A comparison of Mercury's magnetic features to the terrestial features indicates that Mercury's magnetic field is much more distorted than the earth's magnetic field because the planet physically occupies most of its own magnetosphere.

The general characteristics (shape, orientation, etc.) of Mercury's magnetic field are very similar to the earth's magnetic field. The field strength is quite different. Mercury's field strength had been estimated as only 600 gammas. The model predicts the maximum isointensity contour at 500 gammas. The earth's magnetic field strength is approximately 50,000 gammas or 100 times

Mercury's field strength.

Meteorological Phenomena

Atmospheric Pressure

The total atmospheric pressure at Mercury's surface has been estimated at 2×10^{-9} mb, hardly comparable to the earth's standard sea level pressure of 1,013.2 mb. Mercury's surface atmospheric pressure is similar to a reading taken in earth's atmosphere at an altitude of 50 km.

Albedo

Although Mercury is a bright planetary object in either the early morning or late evening sky, its brilliance is more a function of its proximity to the sun than any other factor. Its actual albedo has been measured at 7 percent, compared to 30 percent for the earth and 7 percent for the earth's moon.

Energy Budget

Detailed studies of the planet's thermal balance have not yet been determined. Mariner 10 determined mid-afternoon surface temperatures near Mercury's equator $\simeq 450^{\circ}K$. Temperatures on the dark side were between 89°-$100^{\circ}K$.

Clouds

Clouds, as we know them on earth, do not exist on Mercury but, a haze has occasionally been observed in Mercury's atmosphere. The haze appears as a region of light, whitish clouds. Two possible explanations have been offered for the haze. The first suggestion is that the haze may be dust raised from Mercury's surface. Second, it is possible that intensive but periodic degassing may be occurring on Mercury. The source of the haze clouds may very well be some combination of dust and degassing but only future research will determine which is the more significant process.

Summary

It appears that many of Mercury's characteristics are due to its proximity to the sun. It is precisely this proximity to the sun that thwarts earth-based telescopes in their quest for more information about Mercury. It might almost be said that Mercury's atmosphere is unexciting or dull in comparison to the wide range of processes operating in the other planetary atmospheres. This may, in part, be correct but since Mercury is the planet in closest proximity to the sun, studies of its rarified atmosphere may lead to a better understanding of the harsh interaction between other atmospheres and the solar wind.

References

Broadfoot, A.L. et al., Mercury's atmosphere from Mariner 10: Preliminary Results, Science, 185, 166 ff., 1974.

Chemical & Engineering News, Mercury's atmosphere contains inert gases, Chemical & Engineering News, 52, 15, 1974.

Hartmann, W.K., Significance of the planet Mercury, Sky & Telescope, 51, 307 ff., 1976.

Ness, N.F. et al., Magnetic field of Mercury confirmed, Nature, 255, 205 ff., 1975.

Victorov, A., Atmosphere discovered on Mercury, Space World, K-6-126, 27 ff., 1974.

Whang, Y.C., Magnetospheric magnetic field of Mercury, J. Geophys. Res., 82, 1024 ff., 1977.

CHAPTER III

VENUS

Descriptive Statistics:

Distance from the Sun	0.7 A.U.
Radius	6,050 km
Period of Rotation	243 days
Mass	0.82 of earth
Density	5.23 g cm^{-3}

Introduction

Studies of Venus have always been thwarted by the cloud cover around the planet. Earth-based studies have been able to identify some high altitude atmospheric constituents and determine that changes did occur in the upper cloud patterns. The recent spacecraft encounters have literally parted the clouds and provided man with a wealth of hard scientific data. The picture of Venus that unfolded was a most interesting one. Venus' atmosphere was shown to be one of the most chemically complex and dynamically active planetary atmospheres in the solar system.

The Composition of Venus' Atmosphere

Earth-based studies provided the initial clues to the composition of Venus' atmosphere. Early studies had shown that approximately 20 percent of the incident sunlight is immediately absorbed by two substances, carbon dioxide and sulfur compounds although identification of the latter compounds was impossible prior to the Pioneer probe's results. Spectral studies at high resolution had recorded more than 5,000 absorption lines, many of which had never been observed in any other planetary atmosphere. It was thought that many of the absorption lines could be attributed to carbon dioxide molecules containing rare isotopes such as ^{13}C, ^{17}O, and ^{18}O. The initial discovery of such unusual combinations of isotopes led scientists to suspect that Venus' atmosphere might hold even more surprises than had been anticipated. These suspicions were later verified as the search for more common atmospheric constituents such as oxygen, hydrogen, water, sulfur dioxide, sulfur trioxide, and other suggested chemicals resulted in scientific discoveries analogous to the opening of the mythical "Pandora's Box." In order to maintain a continuity of thought we shall first identify the existence of certain atmospheric constituents and then discuss any interactions or relationships between them.

The gas chromatograph on the Pioneer sounder probe sampled the gaseous constituents at three separate levels, 52, 42, and 22 km, during its descent through the atmosphere, hereafter referred to as sample heights 1, 2, and 3, respectively. Sample height 1 represents data from within the cloud region, sample height 2 from just below the cloud region, and sample height 3 from closer to the surface in what is considered to be the lower atmosphere. The results are presented in Table 3.1. It should be noted that in several columns

TABLE 3.1

Revised atmospheric composition of Venus as measured by the Pioneer Venus sounder probe gas chromatograph (after V. Oyama et al., Science, v.208, p.399, 25 April, 1980. Copyright 1980 by the American Association for the Advancement of Science).

Gas	Flight Sample		
	1	2	3
	Concentration (%)		
CO_2	95.4	95.9	96.4
N_2	4.60	3.54	3.41
H_2O	0.06	0.519	0.135
	Concentration (ppm)		
O_2	?	?	?
CO	32.2	30.2	19.9
Ar	60.5	63.8	67.2
Ne	8	10.6	4.31
SO_2	600	176	185
Altitude (km)	51.6	41.7	21.6
Pressure (bars)	0.698	2.91	17.8

of Table 3.1, the numbers add up to more than 100 percent because of the margin of error factor (±) reported in the original data. The values presented may be viewed as reasonably accurate but subject to more precise definition at some future date.

The high concentration of carbon dioxide (CO_2) in Venus' atmosphere did not come as a surprise to anyone. The presence of CO_2 had been detected from earth. Its presence is consistent with models which seek to explain Venus' high surface temperatures as the product of an exceptionally efficient greenhouse effect. Carbon dioxide absorbs infrared energy very efficiently at most wavelengths. The relative uniform concentration of CO_2 at sample heights 1, 2, and 3 (≈96%) confirmed that it is uniformly distributed throughout the atmosphere. The CO_2 values imply a well mixed atmosphere and suggest that surface temperature patterns may also exhibit some degree of uniformity.

The nitrogen (N_2) content of Venus' atmosphere varied at the different sample heights. The highest concentration (4.6%) was measured at sample height 1 within the clouds. Below the cloud layer at sample heights 2 and 3, N_2 values of ≈3.5 percent were measured. The discrepancy between the values measured at sample height 1, 2, and 3 suggests N_2 may undergo or take part in chemical reactions which convert the gas into some other nitrogen compound. The Soviet Venera 8 probe detected the presence of a nitrogen compound, ammonia (NH_3), in the lower atmosphere. The amount was estimated at between 0.01 and 0.1 percent by volume although there is some uncertainty as to the state of the NH_3. It may be gaseous or it may have dissolved onto particulate matter. This process would be similar to the complex reactions in the earth's stratosphere that produce large ammonium sulfate haze particles. The presence of NH_3 or the formation of haze particles might possibly account for the discrepancy in nitrogen amounts at different sampling heights. The haze particles might also act as a catalyst for large droplet formation and possibly, precipitation. The possibility and significance of droplet formation in Venus' atmosphere will be discussed in a later section of this chapter.

The detection of water on Venus raises a host of questions about the source of the water and its role in the evolution of the planet's atmosphere. The actual concentration of water clearly indicates that the bulk of the water is below the cloud region (Table 3.1) but well above the surface. The highest concentration (0.52%) was detected at the 44 km sampling height. The values leave little doubt that Venus' atmosphere is extremely dry. The total concentration of H_2O in the atmosphere has been estimated at <1 percent.

Earlier probes reported the presence of hydrogen gas in Venus' atmosphere. Scientists suspected that this hydrogen might be the remains of Venus' early water supply. Dissociation of the primordial water supply by sunlight should have produced an atmosphere rich in deuterium (D_2), a heavier, isotopic form of hydrogen. The lighter hydrogen should have escaped to space leaving more of the heavier deuterium isotopes behind. The Mariner 10 probe detected hydrogen but the sought after abundance of deuterium was absent. Based on this finding, it now appears that the source of the hydrogen in the upper atmosphere is most likely the solar wind (as on Mercury) rather than the trace remains of any primordial water supply that may have existed on Venus.

It would still seen reasonable to expect more water on Venus than has been observed. The presence of oxygen, dissociated from carbon dioxide and hydrogen, should combine to form water. It appears, at the present time, as if the available hydrogen combines with other substances, particularly chloride (Cl) to form hydrogen chloride (HCl) however, until a detailed atmospheric chemistry profile is prepared, this may be more conjecture than scientific fact. It remains clear that, for whatever reasons, the atmosphere of Venus is virtually devoid of water.

The presence of oxygen atoms on Venus was expected but the concentrations and altitudinal distribution remain an enigma. Mariner 10's ultraviolet spectrometer detected 10 times more atomic oxygen in Venus' upper atmosphere than Mariner 9 found at similar levels on Mars. Since atomic oxygen readily combines to form O_2, the absence of molecular oxygen (O_2) in the Venusian stratosphere was truly puzzling. Infrared spectra also revealed the presence of carbon monoxide on Venus. It is assumed that the carbon monoxide is formed in the upper atmosphere where carbon dioxide is dissociated by ultraviolet radiation. The inevitable by-product of this reaction is molecular oxygen. The expected ratio of oxygen to carbon monoxide molecules should be 0.5:1.0, particularly since this was the ratio previously discovered on Mars. Actual observations showed that the oxygen content of Venus' atmosphere is actually 50 times less than the amounts on Mars. Two theories have been suggested to explain this deficiency of molecular oxygen. Molecular oxygen may actually be formed as suspected but may also readily combine with some other gas at that level. Or, it is possible that the atomic oxygen created by the dissociation of carbon dioxide is immediately transported downward into the cloud regions where it could chemically combine with other substances, possibly sulfur. If the oxygen is transported downward, then turbulent mixing in the Venusian stratosphere may be of considerable magnitude and responsible for large scale overturning of the atmosphere.

Ultraviolet absorption by molecular oxygen in the earth's stratosphere produces ozone and the ozone layer. Venus does not have an ozone layer. The greatest extinction of ultraviolet energy occurs as the UV wavelengths are absorbed above and in the cloud regions. What remains, penetrates to an altitude of 25-30 km, with very little, if any, penetrating below 25 km. The concentrations of atomic and molecular oxygen are so low in the ionosphere, and notably absent in the Venusian stratosphere, that ultraviolet radiation penetrates much deeper into the atmosphere.

An interpretation error in the Pioneer gas chromatograph data makes it impossible to determine the oxygen content of Venus' atmosphere below 54 km at this time. All evidence suggests a well mixed atmosphere. If so, one would expect the oxygen content values measured at different sampling heights to be similar. While this assumption may be correct, it is conceivable that oxygen amounts at lower altitudes may be depleted because the oxygen becomes incorporated in numerous chemical reactions driven by the high temperatures near and at the planet's surface. Definitive answers await the results from future probes to Venus.

The formation of carbon monoxide (CO) from the photodissociation of CO_2 has already been mentioned. The amounts of CO reported at sample heights 1, 2, and 3 were 32.2, 30.2, and 19.9 ppm, respectively.

The amount of argon, one of the noble gases, increases with decreasing altitude with values of 60.5, 63.8, and 67.2 ppm, respectively. Argon exists in three forms ^{36}Ar and ^{38}Ar (referred to as non-radiogenic argon), and ^{40}Ar (radiogenic argon). Radiogenic argon is produced by the decomposition of potassium (^{40}K) in land rocks. Its presence in the atmosphere is less representative of atmospheric evolutionary processes than it is of the age of the rocks on Venus. Atmospheric (non-radiogenic) argon is 200-300 times more abundant on Venus than on earth. This quantity was totally unexpected and forced scientists to re-think the role of the noble gases. Scientists had hoped to correlate the amounts of argon and the other noble gases with a planet's volatile, reactive elements. This was based on the theory that, as the sun and solar system formed, the temperature decreased with increasing heliocentric distance. Lower temperatures farther away from the sun would have allowed more volatiles and noble gases to be incorporated into the materials which were to become the planets. The large amounts of atmospheric argon on Venus negated these ideas. A revision of these ideas suggests that the temperatures within the pre-solar system nebula did not increase greatly toward the center (the area now occupied by the sun). Instead, the greater gas pressures closer to the center region caused more argon to adsorb onto the primordial planetary materials. If true, this theory would explain the high argon values on Venus. It would also correlate with the lesser amounts of argon found on earth. Mars, being farther removed from the sun has even less argon in its atmosphere than the earth.

The pattern presented by the distribution of argon on Venus, earth and Mars also applies to the concentrations of neon (Ne), and krypton (Kr), other noble gases. Neon amounts at sample heights 1, 2, and 3 were <8.0, 10.6, and 4.3 ppm, respectively (Table 3.1). The best estimates for krypton on Venus range from 0.5 to <1.0 ppm. These data are not formally reported in Table 3.1 because they were derived from data from an instrument that experienced functional problems. It is possible that more krypton exists on Venus but these data are all that are currently available. It is certain that Venus also has more Ne and Kr than either the earth or Mars. The amounts of the remaining noble gases, helium (He) and xenon (Xe), have not been determined for Venus. Logically, we could expect concentrations of He and Xe to be greater than those concentrations in the earth's atmosphere and much greater than those on Mars.

The sulfur cycle on Venus may represent one of the planet's most interesting balances when it is finally understood. The sulfur gases and compounds are intricately linked to Venus' atmospheric chemistry. Scientists studying Venus had suggested sulfur bearing gases as atmospheric constituents based on the yellowish color the planet exhibits on photographic images. The overall composition of our solar system also made it likely that sulfur compounds existed on Venus. More recent ultraviolet absorption studies and probes have confirmed sulfur's presence in the atmosphere. The real questions have yet to be answered. The initial questions relate to the identification of those specific compounds of sulfur and their concentration levels relative to altitude. The more difficult questions raise the problems of understanding sulfur's role in the thermochemistry of the lower atmosphere. It is important to remember that chemical reactions in Venus' upper atmosphere are primarily photochemical processes driven by the sun's energy. Little sunlight reaches the planet's surface so the high temperature regime that exists is the primary driving force for any chemical reactions, hence the use of the term, thermochemistry.

Early research suggested that a photochemical destruction of sulfur bearing gases would occur in the upper atmosphere of Venus and that carbon oxysulfide (COS), was the dominant gas. It was assumed that the photodissociation of carbon dioxide (CO_2) would make carbon monoxide (CO) available to combine with sulfur (S), to form carbon oxysulfide (COS). The COS would in turn be photodissociated producing CO which would combine with an atomic oxygen (O) to form CO_2. The sulfur would combine with various forms of oxygen to produce sulfur oxides: SO_2 and SO_3. These sulfur gases would react with various combinations of hydrogen or hydrogen and oxygen, i.e. H, OH, HO_2, and H_2O to form sulfuric acid, H_2SO_4.

The Pioneer gas chromatograph determined that sulfur dioxide (SO_2), not carbon oxysulfide (COS) was the dominant sulfur bearing gas in Venus' atmosphere. Concentrations of <600, 176, and 185 ppm were detected at sample heights 1, 2, and 3, respectively. The margins of error associated with these SO_2 data are so large that these specific values should only be viewed as reasonable estimates. These data do not prove the earlier suggestions for COS totally wrong. The suggested reactions need only be modified to account for the dominance of sulfur dioxide (SO_2). The end products remain the same.

Therefore, while SO_2 is the dominant sulfur bearing gas on Venus, the presence of some COS is likely, though in quantities significantly less than previously expected.

The gases listed in Table 3.1 represent an initial inventory of the more significant atmospheric constituents of Venus' atmosphere. There is little doubt that many other minor constituents will eventually be identified. More refined analyses of the Pioneer data will also provide estimates of the currently unavailable O_2 content of Venus' atmosphere.

Atmospheric Chemistry

The atmospheric chemical reactions are photochemically driven at and above the clouds and thermochemically driven in the high temperature region of the lower atmosphere. A model depicting the various chemical reactions and pathways is presented in Fig. 3.1.

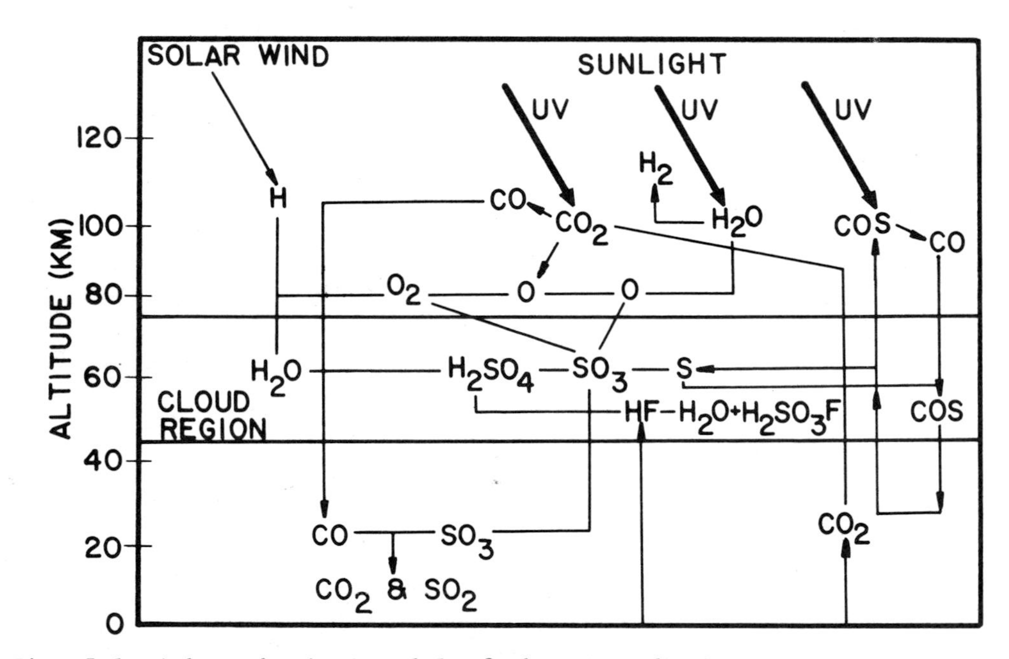

Fig. 3.1. A hypothetical model of the atmospheric chemistry on Venus.

Several initial observations may be made from Fig. 3.1: (1) the atmosphere is deep, (2) it is layered, (3) there is a cloud region (the shaded area in Fig. 3.1), and (4) the presence of numerous acids portrays an image of a planet whose weather and climate are controlled by chemistry rather than by differential heating. Observations 1-3 shall be discussed in the next section of this chapter dealing with the structure of the atmosphere. The presence of unusual chemicals in Venus' atmosphere (observation 4), as previously mentioned, had long been suspected. The atmosphere's yellowish hue and the unusual absorption characteristc at 3-4 μm and 11.2 μm provided the initial clues that many of the atmosphere's absorption characteristics could be attributed to the existence of sulfuric acid. The formation of sulfuric acid appears to be the result of a series of photochemical reactions culminating in the formation of sulfur dioxide (SO_2) and sulfur trioxide (SO_3). Any sulfur trioxide formed in the cooler atmospheric layers would combine with the water vapor there and form a haze layer of sulfuric acid. Similar reactions occur in the earth's polluted urban atmospheres. Concentrations of hydrogen chloride (HCl) have also been observed in Venus' atmosphere. For several years, scientists debated whether sulfuric acid or hydrochloric acid was the dominant acid. This issue appears to have been resolved when it was reported that solutions containing about 75 percent sulfuric acid by weight remain liquid at temperatures of $250 \pm 10^{o}K$. This temperature also exists in the cloud regions of Venus. Polarmetric observations reported a refractive index of 1.44 which is also consistent with H_2SO_4 (at 75-80% concentration). Solutions containing hydrochloric acid less than 32 percent by weight would violate the refractive index results. In order to generate the same effect, stronger solutions requiring 10^4 times more hydrogen chloride vapor would be necessary. Spectroscopic observations show that such concentrations are not present on Venus. Furthermore, absorption at 11.2 μm cannot be attributed to HCl, only H_2SO_4. Based on these findings, we may conclude that the composition of the Venusian clouds is more fully explained by sulfuric acid rather than hydrochloric acid solutions. A more detailed discussion of the cloud region follows in a later section of the chapter.

There is no doubt that small amounts of hydrochloric (HCl) and hydrofluoric (HF) acids are present in Venus' atmosphere. These substances were probably removed from surface rocks by the high temperature regime that exists at the surface. The initial discovery of sulfuric, hydrochloric, and hydrofluric acids came as somewhat of a surprise. As our knowledge of Venus' atmospheric chemistry expands it appears evident that the Venusian atmospheric environment is probably the most reactive one in our solar system. The model in Fig. 3.1 traces pathways whereby carbon dioxide, oxygen, water, sulfur trioxide, hydrogen chloride, and hydrogen fluoride may enter a series of reactions whose end products are sulfuric, hydrochloric, hydrofluoric, and fluorosulfuric acid droplets. These droplets may fall toward Venus' surface as a powerful, corrosive rain capable of intense chemical weathering of the landscape. The actual presence of significant topographic elevations observed by the Pioneer probe suggest that tectonic processes, i.e. mountain building, etc., may occur at faster rates than on earth.

It is particularly obvious that Venus' atmospheric chemistry is extremely complex and not fully understood. The model present in Fig. 3.1 is a first approximation, an initial attempt to understand and explain the basic atmospheric chemistry. It will certainly be modified many times as new data become available.

The Structure of Venus' Atmosphere

Temperature

The temperature profile depicts a general pattern of increasing temperature with decreasing altitude (Fig. 3.2). The profile

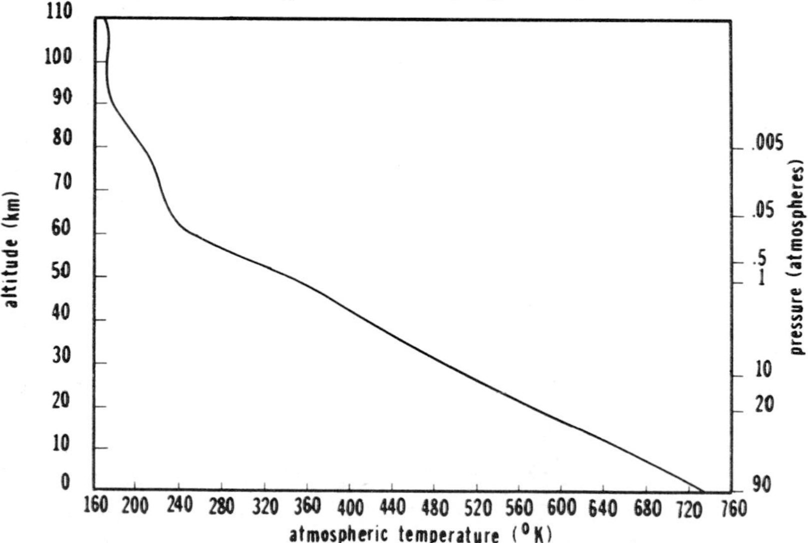

Fig. 3.2. The temperature structure of Venus' atmosphere from 0-110 km. (After A. Seiff et al., Science, v 205, p. 47, 6 July 1979, Copyright 1979 by the American Association for the Advancement of Science.)

represents an accurate picture of the temperature structure of the cloud layer and the lower atmospheric layer. Only a limited portion of the upper layers is represented in Fig. 3.2. The vertical profile shows a surface temperature of $\simeq 750^{o}K$ declining steadily with altitude to $493^{o}K$ at the base of the clouds. There is a distinct slope change at $270^{o}K$ at the base of the upper cloud region (57-58 km). The temperature at the top of the upper cloud layer (68 km) is $230^{o}K$. The temperature decreases more rapidly above this altitude to a nearly isothermal state between 95-110 km with temperature values $\simeq 170^{o}K$. Temperatures at higher levels begin to increase dramatically for the same reasons temperatures in the earth's atmosphere increase with altitude through the thermosphere.

The data from the four Pioneer probes did not report any significant temperature variations below the cloud region in the lower layer. The northern probe did observe temperature $\simeq 25^{o}K$ higher than the other three probes. The temperature profile's slope change in the middle cloud region reveals temperatures warmer by 10°-$20^{o}K$ in the clouds. The warmer temperatures create the thermal gradients resulting in the convective instability. Temperature contrasts between the night probe and the three day probes up to 80 km is $1^{o}K$. The contrasts from 80-100 km is $5^{o}K$ with the day probes reporting warmer temperatures.

The six temperature profiles in Fig. 3.3 suggest a strong latitudinal dependence. The 80 km and 90 km profiles show temperatures increasing poleward by $\simeq 10^{o}K$. The contrast is absent from the 100 km profile. At that altitude, the day side is warmer in the equational latitudes decreasing poleward and continuing to decrease equatorward on the night side of Venus. The temperature difference is $\simeq 5^{o}K$. The 70 km profile represents a transition profile from the upper atmosphere to the cloud region. The 11.5 μm and 50 μm profiles cannot be equated with any set altitude because they represent the cloud top (11.5 μm) and inner cloud regions. Both the 11.5 μm and 50 μm curves reveal profiles whose temperatures increase from the equatorial to middle latitudes then decrease sharply ($\simeq 30^{o}K$) between 60°-80° latitude. The temperature then increases abruptly at the poles. The sharp decrease in temperature (60°-80°) represents the higher and hence colder, collar clouds that ring the poles. The 11.5 μm and 50 μm data show a significant

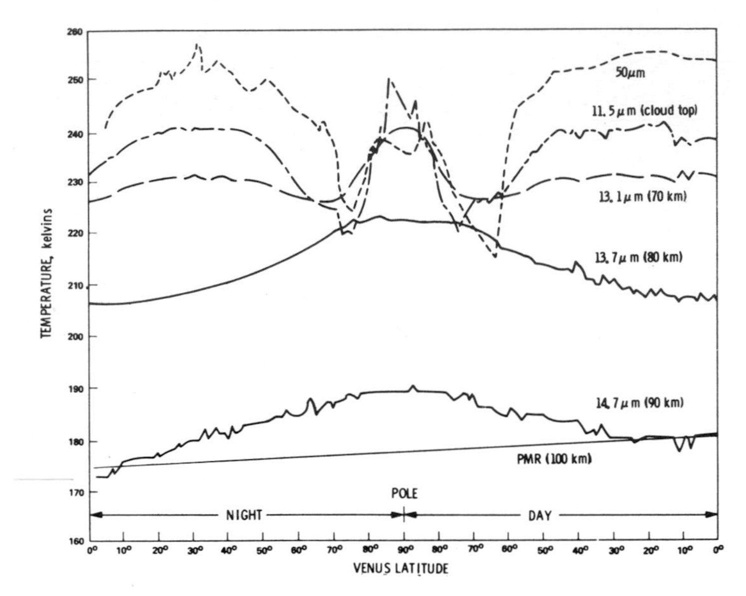

Fig. 3.3. Latitudinal temperature profiles for six altitudinal levels. (Courtesy F.W. Taylor et al., Science, v 203, p. 780, 23 Feb. 1979, Copyright 1979 by the American Association for the Advancement of Science.)

temperature increase from the collar cloud region poleward. Polar temperatures of 240°K (11.5 μm) and 250°K (50 μm) were observed. More recent data reports polar temperatures of 260°K at these latitudes. The implications of these higher temperatures are extremely significant. They suggest that the poles are cloud free and/or suggest the presence of strong subsidence. This subsidence may be associated with Venus' general atmospheric circulation above the polar region. The high temperatures above the poles raise additional problems for scientists attempting to describe a model of Venus' general circulation that is consistent with the temperature and cloud data.

Stability

The temperature structure directly affects the stability of the atmosphere. The stability of the atmosphere in turn determines the propensity for vertical motions. Atmospheric stability is determined by comparing observed temperature/altitude variations, (the lapse rate), to a model plot of temperature/altitude variations known as a dry adiabat. The dry adiabat for Venus' atmosphere should be a line depicting approximately a $9^{o}K\ km^{-1}$ temperature decrease with height (Fig. 3.4). In comparison, the dry adiabatic lapse rate on earth is

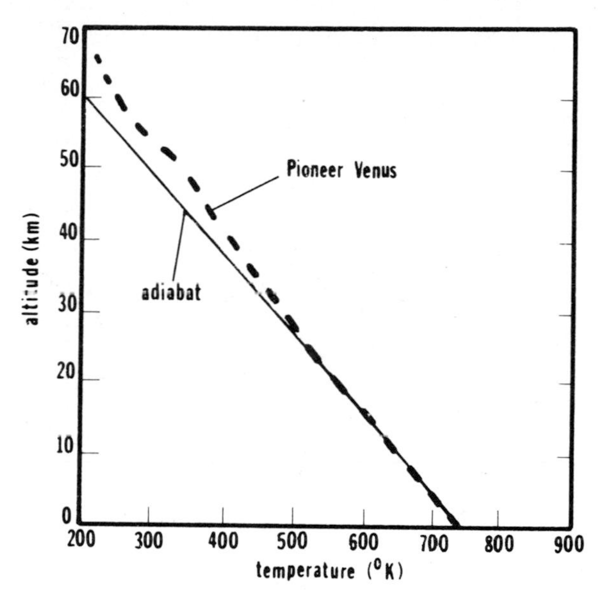

Fig. 3.4. A comparison of the vertical temperature profile with the adiabat. (After A. Seiff et al., Science, v 203, p. 788, 23 Feb. 1979, Copyright 1979 by the American Association for the Advancement of Science.)

approximately $10^{o}K\ km^{-1}$. The atmosphere is stable so long as the temperature plot representing the observed data parallels the dry adiabat is Fig. 3.4. It is obvious that at most levels, Venus' atmosphere is stable and little mixing or overturning occurs. The data show indications of some instability at 20-29 km and 52-56 km however, these minor variations do not appear on the curve in Fig. 3.4 due to the size of the diagram. Seif et al. (1979) have plotted the static stability of the atmophere using an index number approach. The theoretical values of temperature decrease with height (Γ) were subtracted from the observed temperature decrease (dT) with altitude (dz). These were graphed against pressure and altitude scales (Fig. 3.5). Positive values represent stability. The negative values clearly delineate two regions of instability at 52-56 km and 20-29 km. Convective motions must therefore occur in these two layers. The UV absorption region is thought to coincide with the middle cloud region (52-56 km). If so, and it seem highly

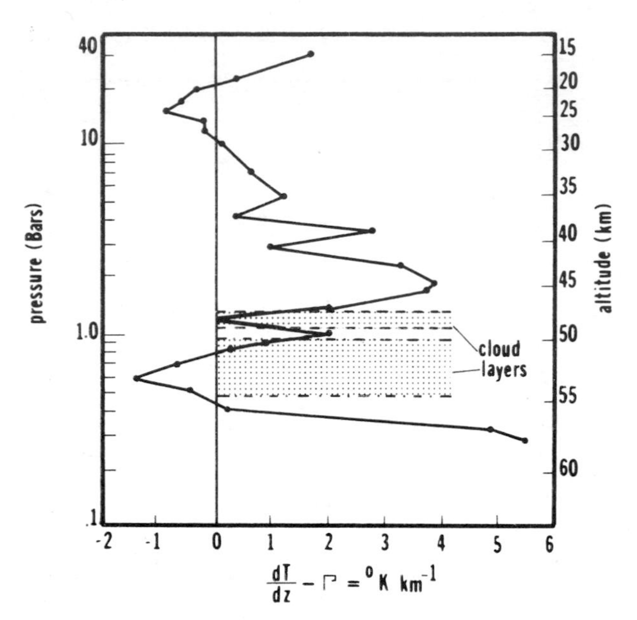

Fig. 3.5. A stability diagram of Venus' atmosphere. Positive values indicate stability, negative values indicate instability. (Courtesy A. Seiff et al., Science, v 205, p. 48, 6 July 1979, Copyright 1979 by the American Association for the Advancement of Science.)

probable that it is, the instability in this region may produce enough vertical convection to explain the convection cell features associated with the four day circulation pattern.

These stability characteristics also exist on the night side of the planet so that the source of the temperature increases necessary to produce the warming and unstable conditions at both altitude zonations mentioned cannot be direct sunlight. The temperature increases probably result from the absorption of outgoing, long wave, infrared energy. It is conceivable that chemical reactions, at least within the middle cloud region, might be a source for some of the thermal energy but, confirmation of this source will await a better understanding of Venus' unique atmospheric chemistry.

Pressure

Pressure data from the various American and Russian probes are in close agreement that pressure on Venus increases with decreasing altitude. The surface pressure, depending on the topography, is in excess of 90 earth atmospheres (Fig. 3.2). The intiial discovery of such high surface pressures startled scientists but now that more is known about Venus, these high values may be explained. Carbon is bound in the biosphere in an organic state on earth. On Venus, the high temperatures have permitted the carbon to remain in the atmosphere in the form of carbon dioxide (CO_2). The result is the 90 atmospheres pressure at the surface of Venus. Theoretical projections show that, if similar temperatures existed on the earth, the oceans would vaporize raising our atmospheric pressure to 200 atmospheres. If all of the earth's carbon was in the form of carbon dioxide, it would raise the surface pressure on earth another 100 atmospheres, to 300 atmospheres. We should appreciate the fact that the earth condensed out at a distance of 150×10^6 km from the sun.

Day/night pressure variations ranging from 4-26 mb at respective altitudes have been observed on Venus. Similarly, meridional pressure differences have also been observed. The meridional differences are, no doubt, pressure gradients associated with the circulations of Venus. These variations may well represent changes in the thermal regimes at select locations. Since detailed pressure maps are non-existent, any analysis of the significance of these meridional and diurnal variations must necessarily await more data.

The Upper Atmosphere

Three distinct atmospheric regions exist on Venus: the region above the clouds/haze layer, the cloud region, and the lower atmosphere beneath the cloud/haze region. This section shall discuss the structure of the upper atmosphere of Venus. The lower layers shall be discussed in the section of this chapter dealing with the "meteorology" of Venus.

There are certain remote similarities between the upper atmospheric layers of Venus and those of the earth. The detection of a turbopause, exobase, and ionopause clearly reveal a zonation of the upper atmospheric regions. The turbopause defines the level where atmospheric gases cease being mixed. The gases begin to form gravitationally separated layers with the lighter gases occupying the uppermost position. The turbopause on Venus is found at an altitude of 140 km while on earth it is found at 100 km.

The exobase is the region from which gas molecules from an

atmosphere escape into space. The exobase defines the bottom of the exosphere, outer space defines the top. The exobase on Venus is at 160 km while the exobase on earth is located at 500 km.

The ionopause defines the upper boundary between the solar wind and the region of charged particles of the upper atmospheric layer known as the ionosphere. The upper ionosphere is dominated by singly charged atomic oxygen (O^+), and is referred to as the F_2 layer. Secondary ions present in the F_2 layer are carbon (C^+), nitrogen (N^+), hydrogen (H^+), and helium (He^+). The peak ion concentration within the ionosphere is termed the F_1 layer. It is located at an altitude ≈150 km. Photochemical reactions produce the ions found at this level. Over 90 percent of the ions are singly charged molecular oxygen (O_2^+). The remaining constituents are nitric oxide (NO^+), carbon monoxide (CO^+), and carbon dioxide (CO_2^+).

The detection of a night side ionosphere on Venus required an explanation. The long night period should eliminate any ions because they are primarily a result of photochemical reactions triggered by sunlight. The presence of ions may be explained if ions created on the day side of the planet are transported to the night side of the planet by the high altitude winds. While transport from the day side may account for the majority of the ions, new evidence suggests that some ionization may be caused by energetic electrons precipitating downward. However, for the present, there is not enought information regarding downward precipitating electrons to explain their contribution to the night time ionosphere.

The characteristics of the night time ionosphere exhibit a high degree of variability. The F_1 and F_2 layers of the night side ionosphere are composed of the same constituent ions as the day side ionosphere. Three possible states of the night time ionosphere have been observed: (1) the ionosphere may be barely detectable with very few ions, (2) the ionosphere may be detectable with high ion concentrations although always fewer than on the day side, and (3) it may be in a transition state associated with changes in the day side ionospere resulting from changes in the solar wind. The night side ionosphere is a more tenuous atmospheric feature because it depends on the day side for its ion supply and is subject to more variations due to processes operating on the day side of the planet.

It is literally impossible to either measure the thickness (depth) of Venus' ionosphere or delimit its altitudinal boundaries because of the solar wind. On earth, the magnetosphere, a strong magnetic field, shields the ionosphere from the effects of the solar wind. Venus' weak magnetism and absence of a magnetosphere permit the solar wind to penetrate farther downward. The solar wind does create a magnetic field in the ionosphere by induction. This field interacts with the solar wind along the top of the ionosphere inhibiting the downward penetration of the solar wind. The interaction produces a bow shock wave between the solar wind and upper atmosphere (Fig. 3.6).

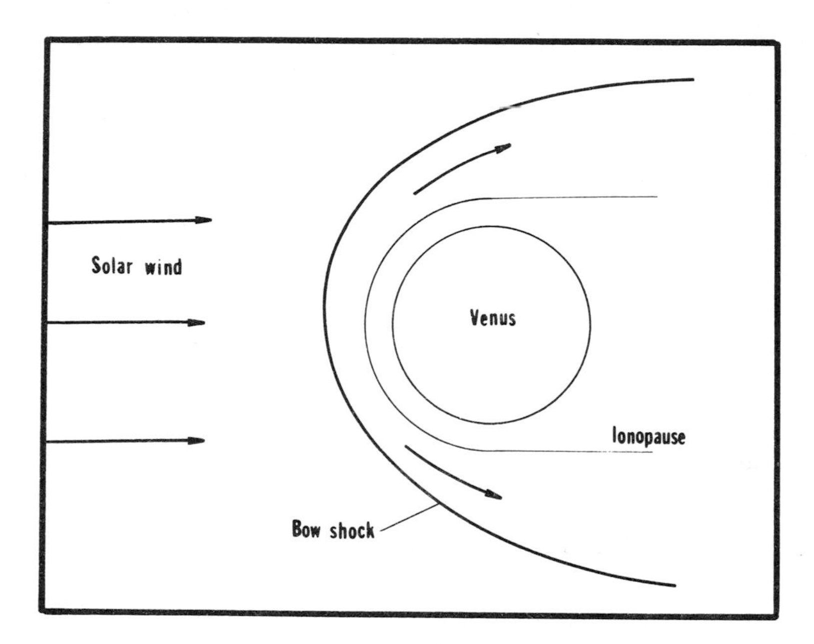

Fig. 3.6. The bow shock and the ionopause where the ionosphere meets and is compressed by the solar wind.

The ionosphere compresses and/or expands in relation to the ram pressure exerted by the solar wind. The Pioneer Orbiter reported that when the speed of the solar wind decreased from 500 km sec^{-1} to 250 km sec^{-1}, the ionopause rose from 250 km to 1,500 km. On one occasion, when a solar flare raised the velocity of the solar wind, within hours the ionopause was compressed down to 250 km. Variations in the speed of the solar wind determine the height of the ionopause and therefore, the thickness of the ionospheric layer on Venus.

Magnetosphere

One of the major objectives of the Pioneer orbiter was to determine if Venus had a magnetic field. Venus' physical size relative to the earth had led scientists to expect to detect the presence of a magnetic field. The magnetic field was expected to be weaker than earth's magnetic field because of Venus' very slow rotation rate of once every 243 earth days. Actual measurements now suggest that Venus does not have an active magnetic field associated with a liquid conducting core. Convection (if any occurs within the planet) is probably not sufficient to generate a magnetic field. Since Venus' geologic structure is still a mystery, the presence of weak convective motions might be due to the current state of Venus' thermal evolution or differences in its geologic structure and/or composition.

It is interesting to speculate on these differences. It had been assumed that Venus' core was similar to the earth's core and therefore, Venus should have had a magnetic field. None has been found. Is it possible that Venus is composed of a different group of chemical compounds? Has its core already cooled? Does the lack of convection preclude plate movements similar to those on earth? These and many other questions merely emphasize the way Venus has presented scientists with the unexpected. Each new fact raises a host of new questions about Venus' evolutionary history.

The absence of a natural magnetic field does not preclude the presence of magnetic effects. Strong magnetic fields have been detected on the night side of the planet. Regions of distinct magnetism with differing field orientations suggest that the magnetism observed is induced by the interaction of the solar wind as it compresses the Venusian ionosphere. A detailed description of Venus' induced magnetism and magnetic properties cannot be made at this time. For the moment, we must be content with what little basic information exists.

Venus' Energy Budget

A detailed knowledge of Venus' energy budget is not available due to the lack of sufficient data. It should be noted, however, that detailed radiation inputs of the earth's energy budget have only been refined to a high degree of accuracy within the past decade.

There is enough data to allow a first approximation of what may occur as sunlight penetrates into the atmosphere. Venus has an albedo of 76 percent, more than twice the earth's albedo of 30 percent, so that only 25 percent of all incident radiation is absorbed by the atmosphere and land surfaces. It is estimated that 24 percent of the incoming energy is absorbed below 75-80 km in the region of the cloud tops by sulfur compounds, mainly at ultraviolet wavelengths. Carbon dioxide is responsible for the absorption characteristics in the near infrared. The remaining energy penetrates the cloud layers but very little reaches ground level.

The net solar energy flux at the outer periphery of Venus' atmosphere is approximately 180 W m^{-2}. The amount of solar energy reaching the surface after atmospheric absorption amounts to between 12-15 W m^{-2}, enough to produce the twilight-like scene of Venus' surface reported by the Soviet Venera probe. The solar energy that does reach Venus' surface probably does so as diffuse light, not direct solar energy. The best estimates suggest that only 2 percent of the solar energy entering Venus' atmosphere penetrates to ground level.

It is extremely important to know the energy pathways (fluxes) in order to construct a model of the planet's energy budget. The downward flux of energy represents the penetration of the solar beam (direct and diffuse) (Fig. 3.7). The upward flux profile represents the thermal infrared energy reradiated after the absorption of the incoming short wave energy. The net flux profile represents the difference between the downward flux (short wave energy) and the upward flux (long wave, infrared energy). The difference between net flux values at different altitude levels represents the quantity of solar energy absorbed between those altitude levels. It is evident from Fig. 3.7 that Venus absorbs more solar energy at every altitudinal level than it radiates out to space. The result is an efficient greenhouse effect and the amazingly high surface temperatures reported by the various space probes.

The efficient greenhouse effect on Venus is primarily a result of infrared absorption by the carbon dioxide in Venus' atmosphere (96%). Surface temperatures as high as 750°K have been reported. This value is considered an accurate measurement of the surface temperature since it has been measured by several different techniques and all

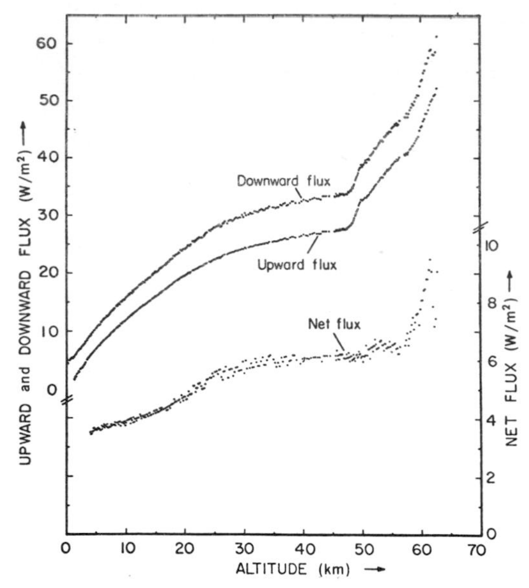

Fig. 3.7. The downward, upward, and net flux of energy in Venus' atmosphere between 0-70 km. (Courtesy M.G. Tomasko et al., Science, v 205, p. 81, 6 July 1979, Copyright 1979 by the American Association for the Advancement of Science.)

are in agreement. However, carbon dioxide alone cannot fully explain the surface temperature conditions on Venus. Carbon dioxide is an excellent absorber of thermal infrared energy but there are windows in the infrared band where CO_2 is transparent to outgoing long wave energy, particularly at wavelengths between 7-9 μm. The high surface temperature could not exist unless these windows were "sealed" by other atmospheric constituents whose characteristics are opaque to those infrared wavelengths that CO_2 is not. Studies of atmospheric constituents have shown that water and sulfur dioxide are present in sufficient quantities to block the radiation windows associated with CO_2. It is therefore possible to explain how such an efficient greenhouse effect can exist on Venus. The greenhouse effect is responsible for the extremely high surface temperatures and the creation of a vast heat reservoir. A schematic model of the energy pathways on Venus is depicted in Fig. 3.8. The heat that is absorbed at the surface and in the carbon dioxide atmosphere is opaque to infrared wavelengths so a strong greenhouse effect exists. Surface temperatures of $750^{\circ}K$ are testimony to its efficiency. Since heat energy cannot escape by direct radiation to space, it is assumed that

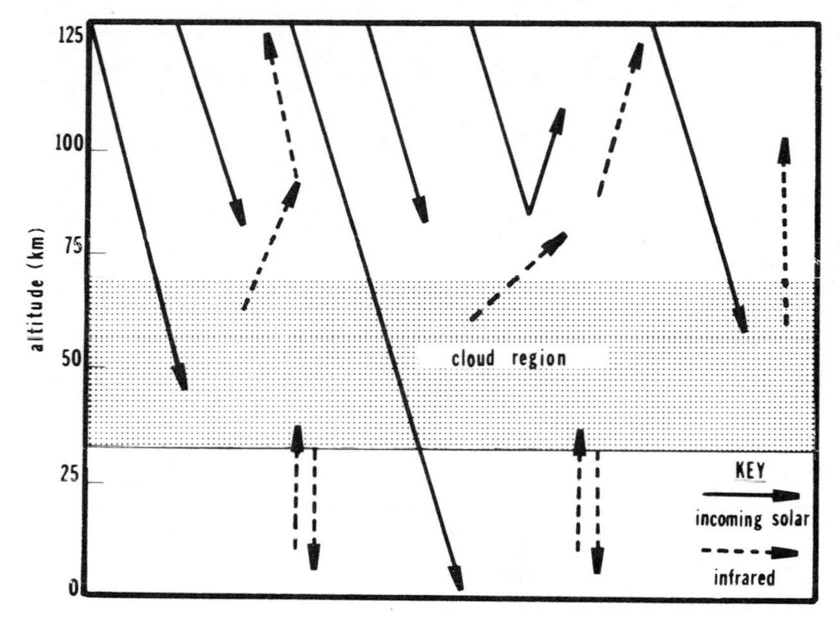

Fig. 3.8. A model of suggested energy pathways in Venus' atmosphere. Solid arrows represent direct solar energy. Dashed arrows represent long wave infrared energy.

strong convective motions transport heat vertically from some level to the cloud tops where it can then be radiated to space. A region of significant instability indicative of convective overturning has been detected below the cloud layers at 20-29 km. Similarly, "convection cell-like structures" have been observed in cloud photographs of Venus taken at ultraviolet wavelengths. While the evidence shows that convection must occur, its importance to energy transport relative to the general circulation of the planet is unknown at this time.

Our limited knowledge of Venus' energy budget is sufficient for us to know that it is quite unlike the earth's energy budget. Venus' atmosphere is heated from the top down. On earth, most solar energy reaching the ground is absorbed and reradiated as infrared energy. The earth's atmosphere is heated from the bottom up. The greenhouse effect is extremely efficient on Venus due to the large quantities of CO_2 along with other minor, but significant, constituents, i.e. H_2O and SO_2. The earth's atmosphere is transparent to the bulk of the infrared energy emitted between 8-12 μm. This energy escapes to space. There are sufficient amounts of CO_2 and H_2O vapor in the earth's atmosphere to absorb enough thermal energy to warm the lower atmosphere on earth and yet maintain a relatively stable balance between incoming solar and outgoing infrared energy. The earth experiences much lower surface temperatures as a result. The uniformity of Venus' cloud cover and energy absorption characteristics produce a uniform thermal environment where surface temperatures between the equatorial and polar regions may vary by only several percent and diurnal temperature variations are no larger than $1^{o}K$. The range of temperatures on earth, both diurnal and latitudinal, are much greater due to variations in cloud cover and the tilt of the earth's axis. Therefore, while Venus is our "sister" planet in terms of diameter and density, it is there that the similarities cease.

General Circulation/Winds

Wind measurements from the Soviet Venera and American Pioneer probes reveal an exceedingly complex atmospheric circulation. The predominant circulation on Venus is zonal (from east to west) at higher altitudes, becoming more meridional (equator to pole) below 90 km. The actual wind speed measurements are a result of vector solutions. If one wind component is easterly (from the east) and the other southerly (from the south), the actual or resulting wind will be a compromise or a southeast wind. This assumes that the individual components are of equal velocity. If not, the prevailing wind will more closely follow the flow of the dominant or higher velocity component. Thus, at altitudes at and above the cloud tops, meridional (southerly) components ≃25 m sec^{-1} and zonal (easterly) components ≃200 m sec^{-1} create observed wind speeds of 100 m sec^{-1}. These winds are part of the four day circulation pattern associated with the cloud patterns and jet streams as they spiral around the planet toward the poles. In contrast to these fast, high altitude winds, virtually calm conditions exist at the planet's surface where wind speeds ranging from 0.4-0.7 m sec^{-1} have been reported by the Soviet Venera probe.

The establishment of a planetary wind field is a by-product of the absorption of incoming solar energy. The earth absorbs the bulk of its incoming energy at ground level and at the sea surface. Thus, the initiation of the earth's general circulation results from temperature and pressure differences created by the differential heating and cooling of land/sea surfaces. Venus absorbs the bulk of its incoming solar energy within and above its cloud layers. The Coriolis effect is virtually negligible because Venus rotates on its axis so slowly. The result is a unique planetary circulation which, in the absence of any Coriolis deflection, takes on the appearance of stacked Hadley Cells.

A proposed model of the general circulation by Seiff (1979) is presented in Fig. 3.9. The predominant zonal flow from east to west

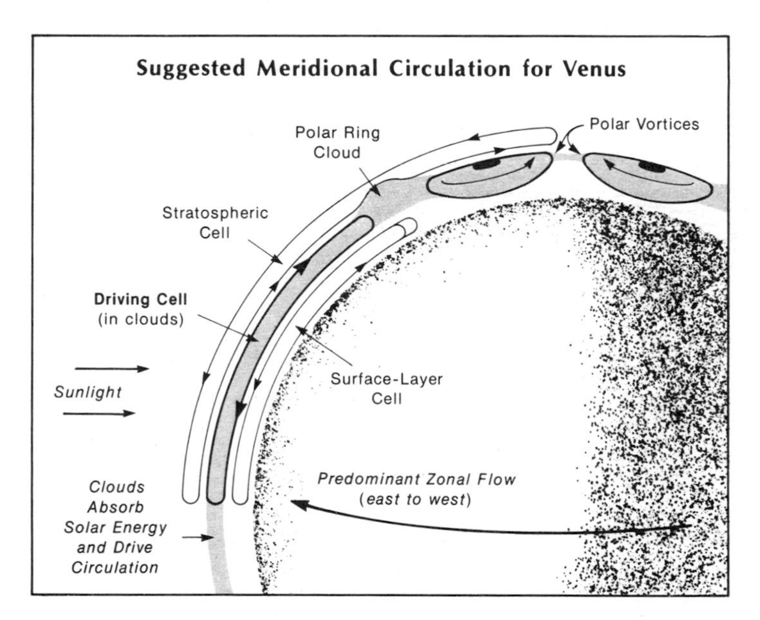

Fig. 3.9. Seiff's cross-sectional model of Venus' general circulation. The lower cell may be further subdivided with the possibility that a slow, directly driven cell exists at or near the surface. (Courtesy A. Seiff et al., NASA Ames Research Center, Moffett Field, California.)

is depicted by the arrow near the equator. The meridional circula-ion of stacked cells is pictured in cross-section. The gray area represents the planet-wide cloud cover. The primary driving cells is within the cloud layers because this region is where the bulk of the sun's energy is absorbed on Venus. The motion of the primary cell (middle) drives the surface-layer cell and the stratospheric cell.

Taylor et al. (1979) derived a model circulation for the 70-100+ km region (Fig. 3.10) based on the Venera and Pioneer vertical wind

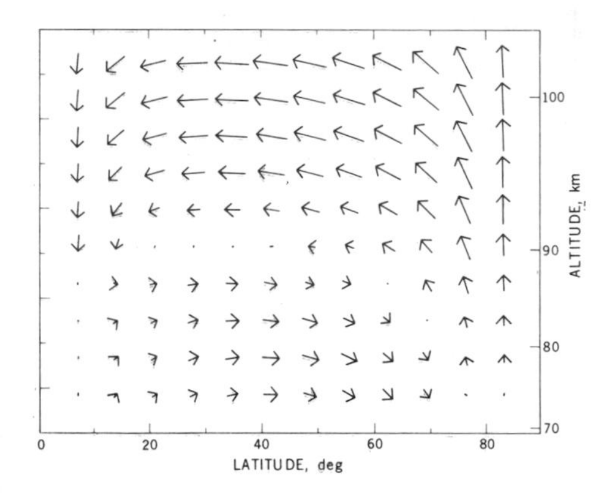

Fig. 3.10. A computer model of the general circulation of Venus from 70-110 km. (Courtesy F.W. Taylor et al., Science, v 205, p. 66, 6 July 1979, Copyright 1979 by the American Association for the Advancement of Science.)

profile data. Taylor's model depicts the direction and strength of the wind vectors in the top of the driving cell (70-90 km) and the return flow in the stratospheric cell (90-100 km) plotted against latitude. Both models appear to agree with the current observational data.

A collar of clouds rings the polar regions at 65°. These are regions of higher and hence colder clouds (Fig. 3.9). The collar clouds are clearly visible as the dark areas in an infrared view of the northern hemisphere (Fig. 3.11). The collar clouds extend up to altitudes of 75 km, and their thickness varies with longitude. The widest section is 2,000 km across and its detailed structure varies with time. Observations have shown that it is widest in phase with solar energy receipt. The collar clouds may be explained either as a region where the normal cloud layers are thicker or as a different, discrete, cloud zone, overlying the main cloud region. Currently, it is felt that the collar clouds are composed of 1-2 μm particles consisting of H_2SO_4 droplets at a concentration of 75 percent.

Two eyes or hot spots (260°K) were discovered adjacent to the pole (Fig. 3.11). These hot spots may be due to subsidence of the main cloud decks or a clearing of the normal clouds. Subsidence would

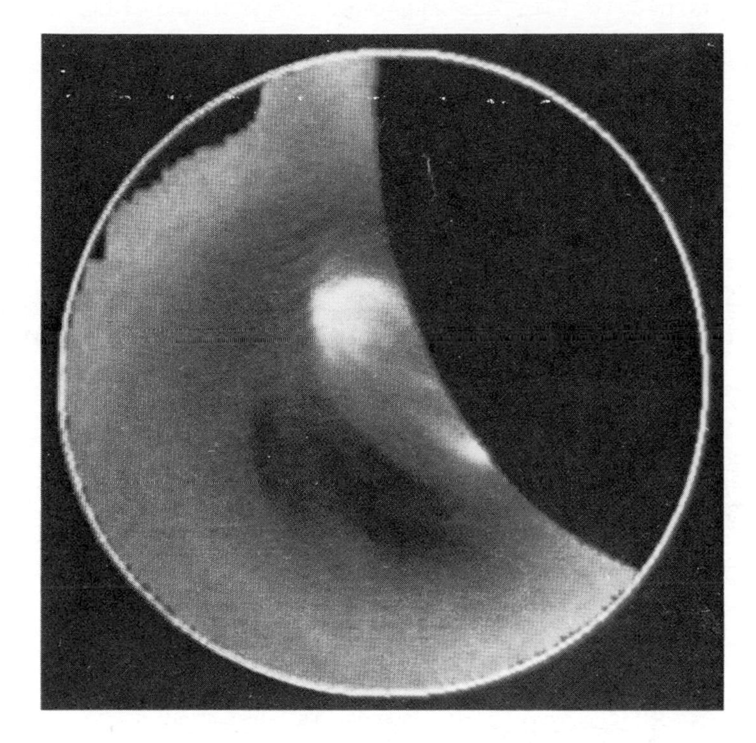

Fig. 3.11. The cold polar collar (darker region) and two polar hot spots (brighter regions). (Courtesy F.W. Taylor et al., Science, v 205, p. 66, 6 July 1979, Copyright 1979 by the American Association for the Advancement of Science.)

create a situation of adiabatic warming of the atmosphere. A clearing of the main cloud region would allow thermal energy from below to radiate upward. Either effect is capable of producing the observed hot regions at altitudes up to 80 km. The existence of these hot spots creates serious problems with the polar circulation suggested by Seiff's model. The polar vortices in Fig. 3.9 cannot explain the observed hotter regions around the pole. Apparently, the model will have to be modified to explain the processes that create and maintain these hot spots.

Four Day Winds

Cloud images from earth based telescopes as well as Mariner 10 and Pioneer Venus Orbiter images reveal a planet whose upper cloud layers are moving in a four day, high speed zonal flow aroung the planet toward the poles. This four day pattern is particularly observable in the low and middle latitude regions. Near the poles, higher wind speeds reduce the pattern to a two day flow. A six day pattern based on spectroscopic measurements is also present. It is possible that both a four day and a six day pattern exist. The ultraviolet absorption studies reveal the upper, four day pattern (Fig. 3.12). Spectroscopic measurements may be "seeing" layers lower in the atmosphere which are moving slower as part of a six day circulation pattern. The negligible wind speeds in the lower atmosphere may eventually be associated with a third, longer period, circulation pattern.

The determination of the zonal circulation is based on the absorption patterns of ultraviolet energy from the sun. The actual absorbing sustance or substances has presented an interesting problem for planetary scientists. The cloud layers are known to consist of various forms of elemental sulfur or H_2SO_4 droplets. The transition region between the upper and middle cloud layers experiences temperatures of 270^oK. At this temperature, an 80 percent H_2SO_4 solution freezes. Modelling studies suggest that the bulk of the ultraviolet energy is absorbed in the main cloud region and that almost all ultraviolet energy is absorbed above 25-30 km. Pollack et al. (1979) suggested five possible ultraviolet absorbers. They are sulfur, meteor particles, metallic particles, nitrogen dioxide (NO_2), and sulfur dioxide (SO_2). The first four possibilities may be rejected because their characteristics do not match the observed characteristics of the ultraviolet absorber. Cloud spectroscopy studies by Stewart et al. (1979) reveal the presence of two regions of absorption near 2.1 μm and 2.8 μm. These absorption characteristics match the characteristics of SO_2 at concentrations thought to exist in Venus' atmosphere. The sulfur dioxide is probably produced by the photodissociation of sulfuric acid cloud droplets and is the most likely substance absorbing the ultraviolet energy.

A sequence of cloud images associated with the four day circulation is presented in Fig. 3.13. The brightness variations may represent different values of the SO_2 mixing ratio, i.e. the actual amount of SO_2 present. The darker bands could conceivably represent regions where more vertical motions carry H_2SO_4 droplets upward. The droplets photodissociate to form more SO_2 in these

Fig. 3.12. An ultraviolet picture of Venus taken by Mariner 10 on February 6, 1974 when it was 720,000 km from Venus. (Courtesy Jet Propulsion Laboratory/NASA.)

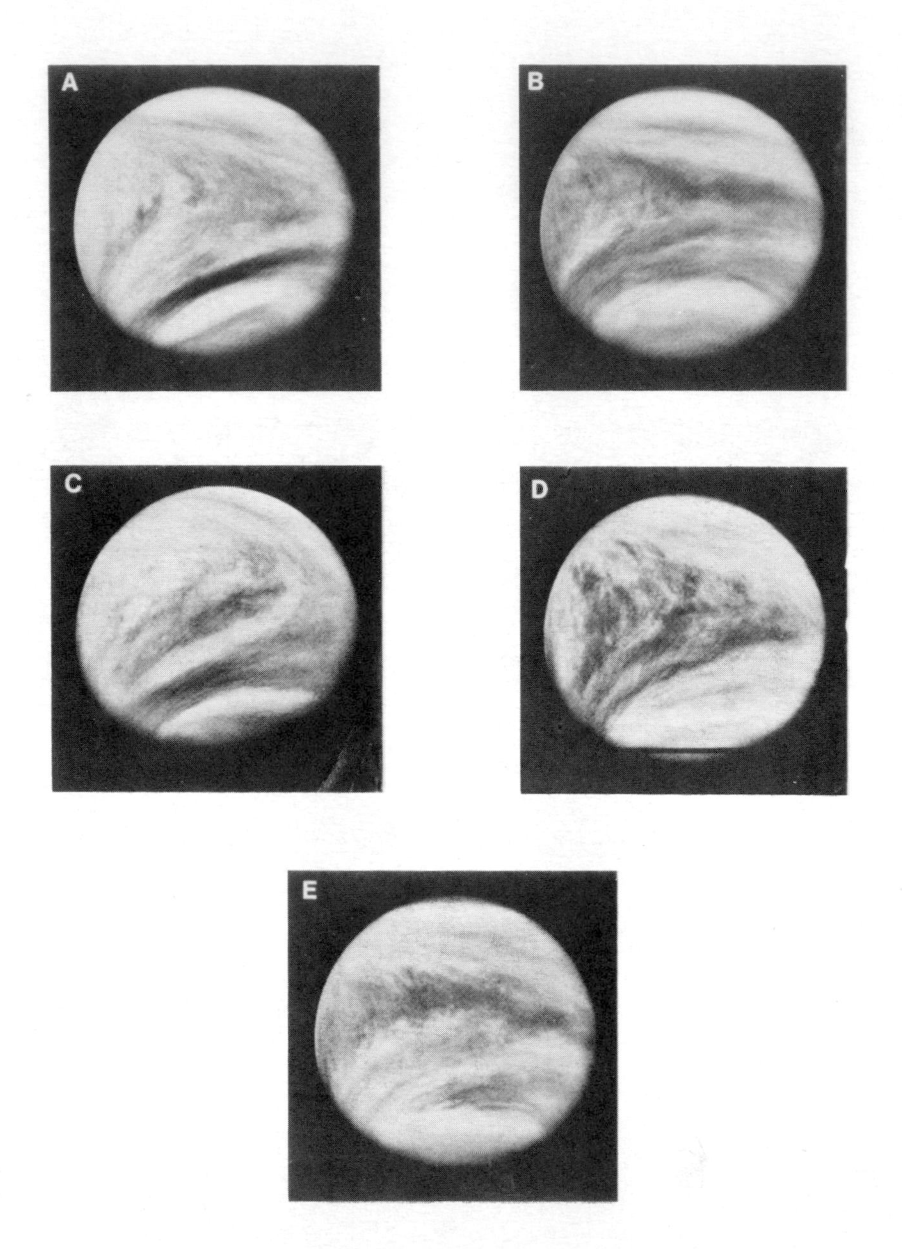

Fig. 3.13. Five ultraviolet images of Venus taken over a six day period illustrating the variability of the cloud features. (Courtesy L.D. Travis et al., Jet Propulsion Laboratory/ NASA.)

regions. More SO_2 at higher altitudes increases the efficiency of ultraviolet absorption hence the darker bands. The pattern of brightness variations often produces a Y shaped pattern visible in Fig. 3.13 A & B. On occasion, the Y shaped pattern does not appear for several weeks. A bow shaped pattern, considered a precursor of the Y shaped pattern, has also been observed (Fig. 3.14). These

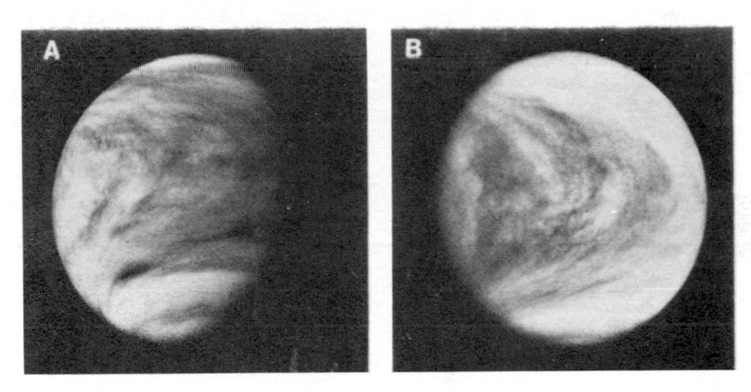

Fig. 3.14. Two ultraviolet images showing the reverse C or bow-shaped pattern in the absence of the Y pattern. (Courtesy L.D. Travis et al., Jet Propulsion Laboratory/NASA.)

bow shapes are also referred to as a "reverse C" pattern. They are typical of those periods when the Y shape is not present. Travis et al. (1979) note that, at higher magnification, the bow shaped markings seem to consist of linked, cellular or polygonal features. These cellular features are consistently found in the equatorial latitudes and have been observed as far poleward as 50°.

Variations in the bright and dark markings appear to be a fundamental part of Venus' global circulation. The life expectancy of the markings is time dependent. The macroscale markings (<1,000 km) in the equatorial and middle latitudes often last for more than four days. Mesoscale markings (100-500 km) only exist for 1.5-4 days.

The study of the Y and bow shaped patterns will eventually permit the development of a detailed general circulation model for Venus. Analysis of the cellular features may provide valuable clues into the nature of local turbulence and/or mixing processes. The results will demonstrate that Venus' atmosphere is an extremely complex entity in terms of its composition and global circulation.

Meteorological Phenomena

Clouds

Historically, the presence of what appeared to be a uniform cloud cover on Venus had thwarted any attempts to learn more about the planet. In recent years, earth based studies made inroads into the veil which surrounds Venus. The data from these studies was often contradictory or, at best, confusing. By 1975, it had been reported that a lower layer of clouds extended from 20-50 km above Venus' surface with a thinner layer, visible only in the ultraviolet portion of the spectrum, above the clouds. Later studies indicated that the main cloud region extended from altitudes of 49-68 km and that a layered structure existed in all but the upper 6 km. These studies also reported that the droplet size in the clouds was restricted to a range of diameters between 0.5 μm and 2.5 μm and that this strongly suggested that these droplets consisted of sulfuric acid. As the body of scientific data concerning Venus grew, all of the data seemed to suggest that the composition of the cloud layers and atmospheric chemistry would reveal a planetary atmosphere unlike anything ever before encountered. The Pioneer Venus probes reached Venus in December, 1978. Their data provided scientists with a first hand look at the most amazing planetary atmosphere in our solar system.

The information concerning Venus' cloud cover is based upon data from the Pioneer Venus probe reports. The data are excellent but limited in that only four probes descended through the Venusian atmosphere. Since the planet is covered by clouds, one should not automatically assume that the results of the probes can be extended to the entire planet. However, Knollenberg and Hunten (1979) suggest that there are several reasons why we could assume the probe's results extend planet-wide. First, telescope based observations reveal that the clouds form a relatively homogeneous layer around Venus. This statement certainly could not be made for the earth whose cloud patterns, when viewed from space, reveal a multitude of diverse cloud forms associated with different atmospheric processes such as convection, fronts, etc. Second, the observed temperatures in Venus' lower atmosphere reveal a lack of variation both locally and geographically with latitude across the planet's surface. On earth, strong temperature variations induce pressure variations which produce winds, moisture (latent heat) transfer, and local weather phenomena. The climatic regions of earth are the result of such surface temperature variations. Therefore, the relatively uniform surface temperatures on Venus permit the assumption of planet-wide cloud uniformity. Third, on earth, cloud particles are a result of the phase variations of water (ice crystals, water droplets, water vapor) associated with temperature variations in the atmosphere. It appears that chemical reactions are responsible for the formation of cloud particles on Venus. Since the thermal regime of the atmosphere appears relatively homogeneous, the reactions that produce the cloud particles probably exist planet-wide also.

Pioneer Venus released four probes (day probe, night probe, north probe, and sounder) to study the structure and cloud layers of Venus' atmosphere (Fig. 3.15). Their combined results, have provided the

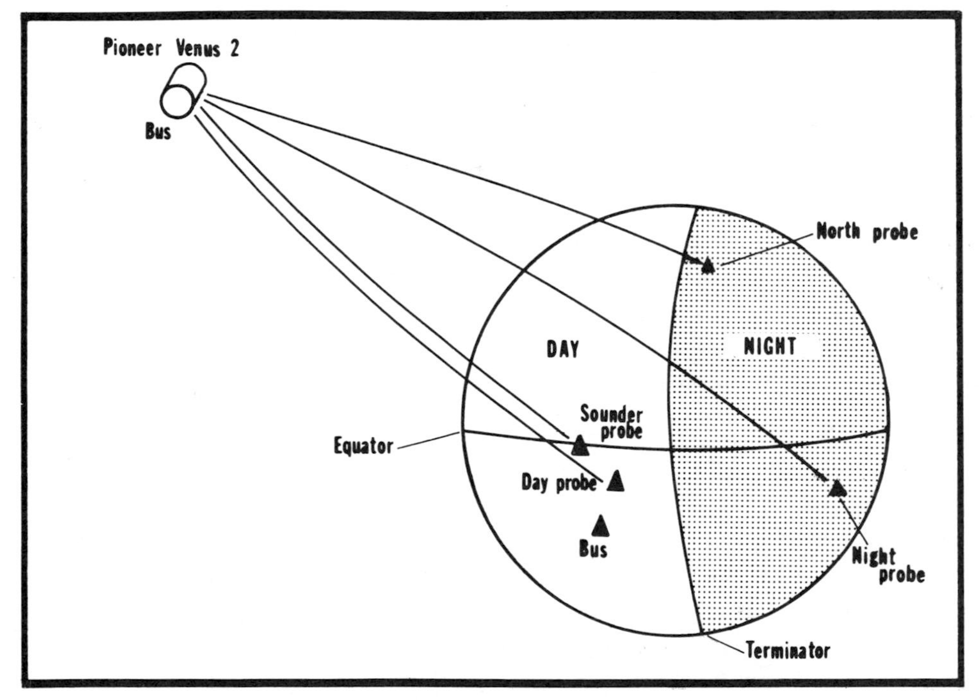

Fig. 3.15. The locations of the various probes launched from the Pioneer Venus 2 into Venus' atmosphere.

first detailed look at the vertical structure and cloud layers of Venus. Knollenberg and Hunten (1979) reported the identification of four cloud regions based on the results of the cloud particle size spectrometer. They (Knollenberg and Hunten) labeled these the upper cloud region, the middle cloud region, the lower cloud region, and the lower, thin haze region found at altitudinal heights of 58-68 km, 52-58 km, 48-52 km, and 31-48 km, respectively. These correspond to the four regions identified by the backscattering nephelometer's results and reported by Ragent and Blamont (1979) as Region D through Region A, respectively. There are preliminary data which suggest a fifth region exists above 68 km. There is also the possibility that a sixth may eventually be recognized as an integral part of the series of atmospheric cloud/haze layers. All that is currently known is that Pioneer Venus data suggest a fifth region above 68 km consisting of a planet-wide haze with particle diameters <0.5 μm. Their characteristics indicate that they may consist of droplets of H_2SO_4 at a 75 percent concentration. Whether this region is a "layer" in the sense used to report the main cloud groups remains to be seen. The Russian Venera probes reported strong indication of another layer below 48 km. This "layer" was not confirmed by the Pioneer Venus probes which reported only a lower thin haze region between 48-31 km. Resolution of the descrepancy between the Pioneer and Venera data is not likely at the present time.

The layered structure of the clouds plotted with the altitudinal zonations is presented in Fig. 3.16. The upper cloud region (Region D, 58-68 km) consists of small particles ranging from 0.6-4.0 μm diameter. The inset bar graph in Fig. 3.16 shows that the bulk of these small particles have diameters of ≃1 μm. These have been designated mode 1 particles. The mode is a term used to designate that class or group in a sample that is greatest in number. The concentration (number density) of these particles ranges from 300-400 cm^{-3}. These cloud particles, while abundant, are not very dense, hence the mass loading (mass per unit volume) value is only 1 mg m^{-3}. A more detailed analysis of the particle size data for the upper cloud region shows a bimodal distribution. Most of the upper cloud layer consists of mode 1 particle sizes but near the botton of the upper cloud layer, mode 2 particles with diameters ≃3 μm occur. Since mode 2 particles are also representative of the middle cloud region, their presence at the lower elevations of the upper cloud region does not present a problem as this would be a transition region anyway.

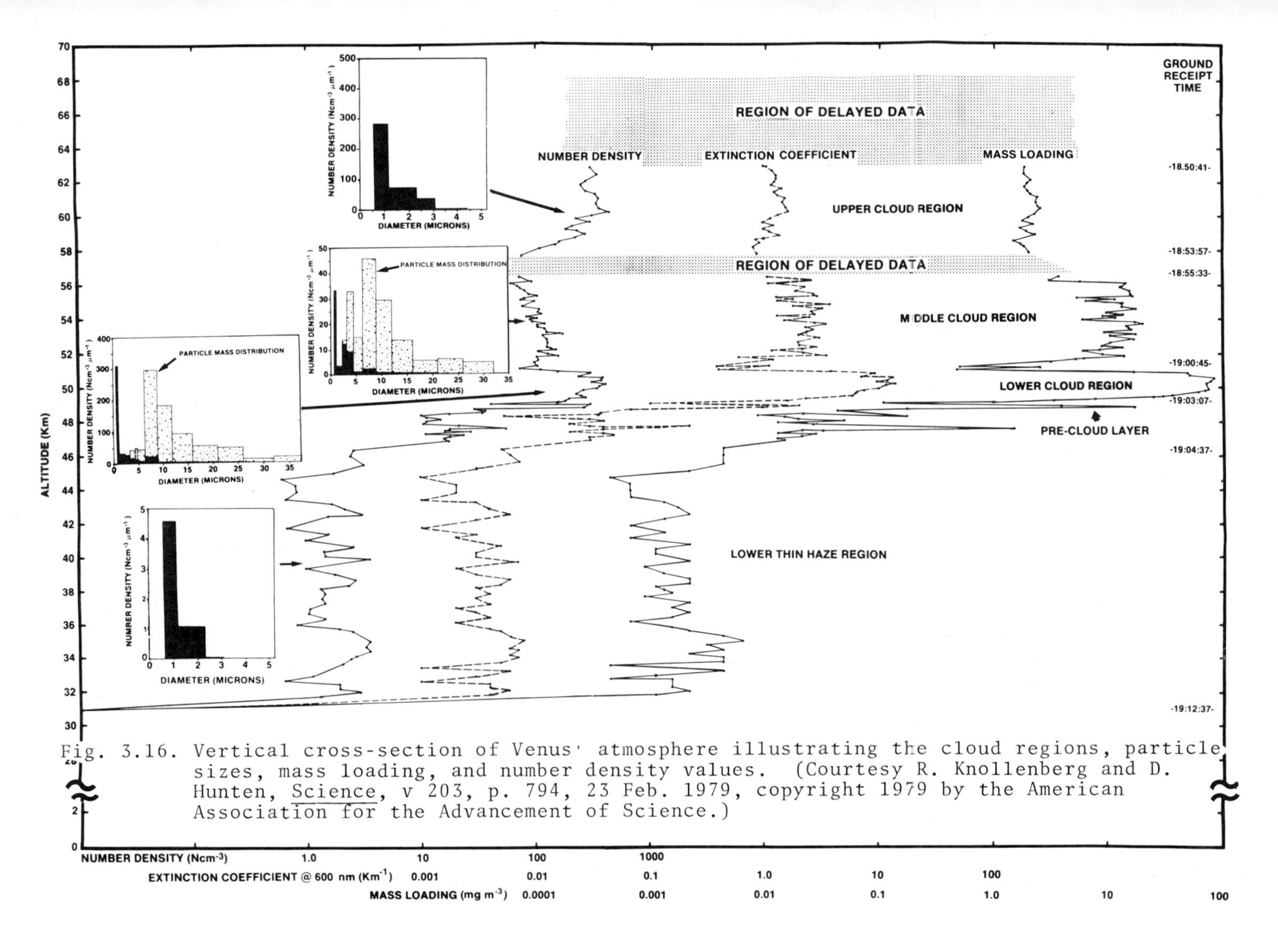

Fig. 3.16. Vertical cross-section of Venus' atmosphere illustrating the cloud regions, particle sizes, mass loading, and number density values. (Courtesy R. Knollenberg and D. Hunten, Science, v 203, p. 794, 23 Feb. 1979, copyright 1979 by the American Association for the Advancement of Science.)

The composition of the mode 1 particles is not definite. Knollenberg and Hunten (1979) first suggested these small diameter particles might be sulfur or H_2SO_4 droplets. Index of refraction data favor H_2SO_4 droplets. Ultraviolet absorption data suggests more elemental sulfur. In a more recent paper, they (knollenberg and Hunten) term these mode 1 particles aerosol debris, suggesting that this aerosol debris is composed of both volatile and involatile particles. It would seem likely that some form of sulfur is present along with other particles and that, depending on the altitude and presence of other compounds, the sulfur may exist as elemental sulfur, H_2SO_4 droplets, or both.

The middle cloud region (Region C, 52-58 km) consists of particles distributed trimodally. There are mode 1, mode 2, and now, mode 3 (≈8 μm) particles (bar graph, Fig. 3.16). Mode 3 particles are larger, ranging in diameter from 5-35 μm. The concentration (number density) of particles is only 100 cm^{-3}, three to four times less than the upper cloud layer. The larger particles do represent a fivefold increase in mass loading to 5 mg m^{-3}. The trimodal size distribution suggests that aerosol debris, elemental sulfur in needlelike shapes, and H_2SO_4 droplets at concentrations of 80-85 percent may all coexist in the middle cloud region.

The lower cloud region (Region B, 48-52 km) is the most dense. Mass loading values range from 10-80 mg m^{-3}. All three particle size modes are found in the lower cloud layer (Fig. 3.17). Mode 3

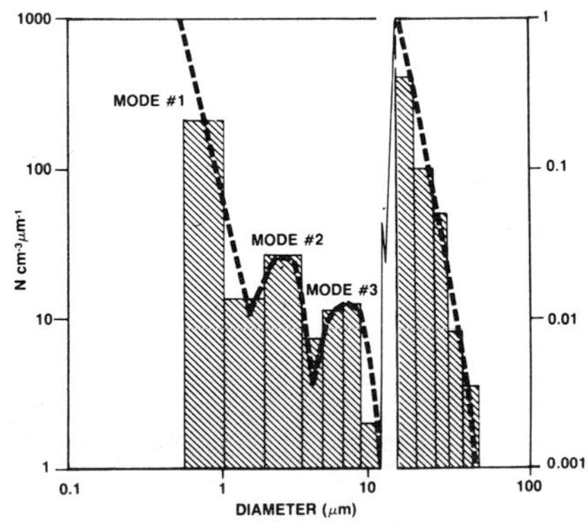

Fig. 3.17. The mean particle size distribution in the lower cloud region. (Courtesy R. Knollenberg and D. Hunten, Science, v 205, p. 73, 6 July 1979, Copyright 1979 by the American Association for the Advancement of Science.)

sizes dominate the lower cloud region and are responsible for the dramatic density (mass loading) increase in this layer. Near the bottom of the lower cloud layer, a separate thin layer appears to exist. In this "pre-cloud" layer, mode 3 particles disappear leaving only the smallest mode 1 particles. The lower cloud and "pre-cloud" layer together account for one half of the total cloud mass on Venus. The size distribution suggests the presence of H_2SO_4 droplets. These probably formed as mode 2 droplets coalesced into larger, mode 3 size droplets. The types of particles, mass loading values, composition, and planet-wide distribution make the lower cloud layer vital to any explanation of the planetary cloud distribution on Venus.

The lower thin haze region (Region A, 31-48 km) consists of a very diffuse region of thin haze particles from 1-10 cm^{-3}. The particle diameters appear to be 1 μm in size with the zone between 45-47 km the most dense portion of the region. The combination of very small particles and such low concentrations results in an atmospheric layer with almost negligible mass.

The general information along with the corresponding temperature values within each layer are summarized in Table 3.2. The altitude

Mode	Modal diameter (μm)	Size range (μm)	Cloud region	Idenity	Concentration (cm^{-3}) U	M	L	H
1		? to 1.5	U,M L,H	S, various aerosols	150	25	100	10
2	2 to 3	1.5 to 5	U,M,L	H_2SO_4, inclusions or impurities	50	50	50	
3		7 to 8	M,L	M:S?, H_2SO_4? L: H_2SO_4		10	30	

TABLE 3.2

A summary of Venus' cloud characteristics (after R. Knollenberg and D. Hunten, *Science*, v 205, p.72, 6 July, 1978. Copyright 1979 by the American Association for the Advancement of Science.)

zones and particle size distribution are now known but problems remain. The composition of the upper cloud layer is still uncertain. Is it aerosol debris or, as its ultraviolet absorption characteristics indicate, elemental sulfur? Logically, both aerosol debris and elemental sulfur probably coexist in the upper cloud layer. Sulfuric acid droplets varying less than 20 percent in diameter throughout 5 km of vertical extent are the primary component of the middle cloud layer. The bulk of the lower cloud also appears to consist of concentrated sulfuric acid droplets but here, their sizes are distributed between mode 2 and mode 3. The probability that mode 2 droplets are coalescing to form larger mode 3 droplets has already been mentioned. It is also very likely that, at higher temperatures near the lower portion of the lower cloud layer, some liquid is evaporated from the mode 3 droplets reducing them to mode 2 sizes again. The coalescence of smaller cloud droplets to produce larger, precipitation size droplets is one of the two processes which produce precipitation on earth. Similarly, the evaporation of falling raindrops is commonplace on earth. It is observed as a veil of mist falling from the base of a cloud, although the droplets never reach the earth's surface. This is termed virga on earth. It does raise the question of whether Venus experiences just virga or a concentrated acid precipitation.

Precipitation

The idea that acidic raindrops may precipitate from some or all of Venus' cloud regions is a logical outgrowth from the information presented about the nature and extent of the clouds themselves. Yet we must dispel the notion that these clouds are, in any way, similar to those naturally occurring clouds on earth. Clouds on earth result from the cooling of rising air. They either condense out liquid water or sublimate ice crystals from water vapor in the air. The cloud particles on Venus would be more like the brown, smog clouds which persist above many of our large urban areas. Knollenberg and Hunten (1979) differentiate between these anthropogenic, smog clouds and naturally occurring clouds on the basis of two parameters. First, smog clouds are chemically created clouds, a by-product of the venting of the waste products of our technological processes into the earth's atmosphere. The clouds of Venus are chemically created from gases, solids, and vapors, naturally existing products of a complex planetary atmosphere. Second, chemically created clouds are difficult to destroy often having life expectancies of many days to weeks. On earth, the smog particles are eventually scavenged by precipitation processes and the chemicals and particulates are transported down to the earth's surface. There does not appear to be a natural scavenging process in the Venusian atmosphere. Particles may move from layer to layer but apparently are not transported to the planet's surface. The clouds of Venus envelop the planet and are extremely persistent as a result. We must therefore qualify our ideas concerning precipitation when we discuss it (precipitation) in relation to the unique atmospheric chemistry of Venus.

The discovery of H_2SO_4 in Venus' atmosphere was followed, quite logically, by theories of corrosive, acid rains. Knowledge of the particle concentrations that increased with depth into the cloud cover suggested that small H_2SO_4 droplets buoyed up by turbulence probably coalesced into larger, heavier H_2SO_4 drops which would then fall due to gravitational attraction. Young and Young (1975) suggested that as H_2SO_4 droplets decended through warmer layers below, some of the droplet's water would evaporate. The H_2SO_4 droplet would become a more concentrated acid solution. These concentrated droplets could, in turn, react with some of the hydrogen fluoride in the lower atmosphere. The result would be fluorosulfuric acid droplets, capable of dissolving mercury, sulfur, lead, tin, and most types of known rock materials. It goes without saying that if Young and Young are correct in their explanation of droplet formation and precipitation, the rain reaching the surface of Venus should be corrosive enough to dissolve any topographic features. Imagine, a planet whose surface might be as smooth as a bowling ball. This, of course, does not seem to be the case as the Venus Pioneer mapper has already reported the existence of both smooth, plateau-like regions and rugged topographic features including volcanoes and mountainous ridges. Knollenberg and Hunten's (1979) version of what may occur is based on more recent information from the Pioneer Venus probes. They suggest that photochemical reactions are a source of the sulfur and/or H_2SO_4 at the upper limits of the middle cloud region ($\simeq$57-58 km). The photochemical processes would be driven by the absorption of ultraviolet energy from the sun. The large sulfur particles would slowly descend into the boundary region between the upper cloud and middle cloud region. Higher temperatures would evaporate the droplets as they continue down toward the transition zone between the middle and lower cloud region. Higher temperatures near the 48 km level would then melt the residues forming the lower haze particles.

The particles with diameters >30 μm in the middle and lower cloud regions grow by the coalescence process, doubling their diameters to sizes large enough to precipitate downward toward the surface. The modal distribution of particle diameters seems to verify the feasibility of Knollenberg and Hunten's hypothesis. The mode 1 particles are distributed throughout all cloud layers and have previously been characterized as H_2SO_4, sulfur particles, and various aerosol debris. The concentration of H_2SO_4 droplets of mode 2 size present in the middle cloud region decreases near the botton of the middle cloud region. These mode 2 particles largely disappear, apparently coalescing into larger, mode 3 sizes whose concentration increases in the lower cloud region. Coalescence may not be the only process aiding the growth of droplet sizes. As liquid evaporates from the droplets, their diameter decreases and acidity increases. They may fall into the lower cloud region, a zone with more H_2O and possibly, SO_2 or SO_3. The droplets would experience renewed growth and their diameters would increase. This would explain the observed density and maintenance of the lower cloud layer. Precipitation may occur but neither scavenges cloud particles nor reaches the planet's surface. The clouds persist as a result.

The precipitation on Venus would be more analogous to virga on earth. Droplets which do manage to achieve diameters sufficient to fall towards the surface of Venus encounter progressively higher temperatures and a major region of instability and overturning at 20-29 km. Temperatures of 575°-550°K would evaporate the volatile substances. The droplet's nucleus, most likely a solid, aerosol particle of smaller diameter, would then be returned to higher altitudes by convective overturning or turbulence.

The precipitation, virga, and cloud regions are unique zones of chemical activity. Photochemical reactions dissociate particles of the upper boundaries. These settle through the layers but later are recycled back. During the settling, H_2O and SO_3 condense around these nuclei forming H_2SO_4 droplets. The heavier droplets sink downward, collide, coalesce and eventually evaporate at higher temperatures. The aerosols are vapors returned aloft to begin the cycle anew. The longevity of these chemically created clouds explains why the surface of Venus remains hidden beneath a most persistent cloud cover.

Visibility

Visibility is a function of particle diameter and particle concentration with the cloud regions. Horizontal visibility in the upper, middle, and lower cloud layers has been estimated at 6 km, 1.6 km, and ≈0 km, respectively. The two higher layers are more similar to dense hazes. The negligible visibility in the lower cloud layer results from the dense concentration (400 cm^{-3}) of large particle sizes. The lower cloud layer is similar to a stratus cloud on earth. Visibility at an altitude of 10 km has been estimated at ≈12 km, while at the surface (0 km), visibility is ≈3 km. The atmosphere at 10 km is very red in color and surface illumination is similar to a dark, rainy day on earth. The outline of the sun is not visible and a "red murk" prevails, according to the Soviet Venera probe.

Lightning

The occurence of lightning discharges on Venus was first reported by the Soviet's Venera 12 probe as it descended toward a landing on the sunlit side in Decêmber, 1978. The first lightning observation occurred when the probe was 10 km above the surface and numerous events were recorded during the probe's brief 110 minutes of operation from the planet's surface. The Soviets reported that the probe descended through or near a thunderstorm. The probe must have carried some type of acoustic measuring device as they also reported a sound reverberation, presumably thunder, that resounded for fifteen minutes.

Several weeks after the Soviet observations, the Pioneer Venus Orbiter's periapsis (low point in its orbital path) passed well into Venus' night side ionosphere. Data from the electric field detector began recording very strong, impulse signals indicative of lightning discharges (Fig. 3.18). The spikes in the electric field measurements (a to d) representing lightning discharges are particularly evident at the lower frequencies of 100 Hz and 730 Hz. The spikes in the curves coincide with increases in electron density (e) which would be a prerequisite for lightning discharges. Clearly, the altitude of the spacecraft is significant in its ability to detect the lightning

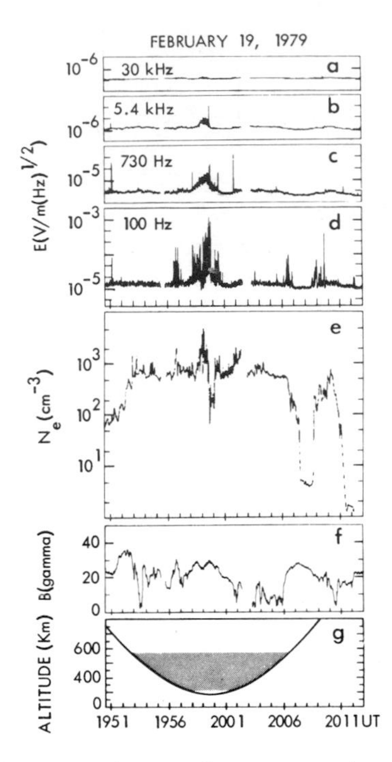

Fig. 3.18. Data from Venus' ionosphere: a-d represent electric fields, e, electron density, f, magnitude of the magnetic field, and g, spacecraft altitude relative to the ionosphere (shaded area). (Courtesy W.W.L. Taylor et al., Science, v 205, p. 113, 6 July 1979, Copyright 1979 by the American Association for the Advancement of Science.)

discharges. In inset (g), the altitude of the orbiter is plotted against time. The shaded area represents the orbiter's penetration into the ionosphere with its periapsis point at 1959 U.T. The majority of the discharges were recorded between 1959 and 2001 U.T. There is little doubt that these events recorded as spikes in Fig. 3.18 are caused by lightning. They are intense, impulsive, occur below the ionosphere and have signals, characteristic of lightning generated wave phenomena as it propagates from lower altitudes into the ionosphere. Lightning has not been observed or detected from the day side of Venus. There are two possible explanations for the absence of day side lightning impulse measurements. It may indicate that Venusian lightning discharges only occur at night or that the daytime ionosphere is somehow strengthened or distorted so that it is opaque to waves propagating upward from lightning discharges in the lower atmospheric layers.

Since the Venera 12 probe reported lightning and thunder from the day side of the planet, the latter of the two explanations seems more plausible.

The observations of lightning and thunder suggest that thunderstorms might exist also. The existing evidence does not support the formation of towering earth-like cumulonimbi with their anvil tops. The Venusian atmosphere below 15 km, between 29-50 km and above 56 km appears stablely stratified. Two regions of instability do exist at altitudes of 20-29 km and 50-56 km. The levels of instability and atmospheric processes necessary to trigger strong vertical updrafts and thunderstorm formation are highly unlikely on Venus.

The absence of cumulonimbi suggests that other processes, most likely chemical, are responsible for the buildup of electrical charge potentials sufficient to produce lightning discharges. Thuderstorms on earth fulfill this role by segregating negative charges in the base of the cloud and positive charges near the top of the cloud, even though the processes producing the charge distribution are not fully understood. The electrical potential gradient prior to a lightning discharge may reach 300 volts m^{-1}. The electrical potential gradient at the extremities of a lightning bolt immediately before discharging may reach hundreds of millions of volts. Perhaps multiple chemical reactions involving sulfuric acid, an excellent electrolyte, result in the segregation of charges in Venus' atmosphere. If so, the region of sulfuric acid droplets would have to extend down closer to the planet's surface. Perhaps "acid rain" does, at least on occasion, reach Venus' surface.

Night Airglow

An airglow associated with the night side atmosphere of Venus is detectable in the ultraviolet portion of the spectrum. This airglow should not be confused with the ultraviolet absorption (by H_2SO_4) of that portion of the sun's energy reaching Venus. The airglow is an emission effect with the strongest emissions occurring at altitudes between 100-120 km. The ultraviolet emissions occur at specific wavelengths ranging from 1.9-2.7 nm.

The source for the night airglow's emission appears to be the chemical recombination of nitrogen (N) and oxygen (O) atoms. These atoms are products of photodissociation of nitrogen and oxygen compounds on the day side of the planet. The high speed, upper winds transport the nitrogen and oxygen atoms to the night side of Venus where they recombine, radiating small amounts of ultraviolet energy in the process. Similar airglows have been observed for the earth and Mars. The airglow on earth and Mars is produced by the same process responsible for the airglow on Venus.

The spatial and temporal distribution of the airglow show that it is brightest on the night side above the equatorial regions. The glow has also exhibited a tendency to be brighter along the night side meridian 30° beyond the antisolar meridian. Temporal observations show that the airglow's brightness varies irregularly with time suggesting irregular changes in the processes operating in Venus' upper atmosphere. These seemingly irregular characteristics may provide a mechanism whereby scientists may be able to "trace" night side upper atmospheric processes and understand the degree to which the winds transport day side products of chemical and/or photochemical reactions to the night side of the planet.

Summary

The atmosphere of Venus is an extremely complex entity. Its composition is primarily carbon dioxide, however, its photochemistry and thermochemistry are unique to the solar system. Upper atmospheric reactions are driven by the absorbed ultraviolet energy. The greenhouse effect and $750^{\circ}K$ surface temperatures may drive a series of thermochemical surface or near surface reactions which we may only speculate about at this time. The atmospheric products of sulfuric hydrochloric and fluorosulfuric acids create extremely corrosive droplets in Venus' atmosphere.

The temperature and pressure structure reflect expected trends of increasing values with decreasing altitude. The 90 atmosphere surface pressure is attributed to the pressure of carbon in the form of carbon dioxide in the atmosphere. On earth, the carbon has chemically combined into the lithosphere and biosphere.

The cloud layers reveal a distinct zonation with each layer consisting of its own constituent particle sizes. The uniform zonation suggests a general circulation similar to three stacked Hadley Cells in each hemisphere. The middle cell appears to be the "driving" cell since most solar energy is also absorbed at these altitudes. A problem of integrating the cold, polar collar clouds and polar hot spots into the model of the general circulation still exists.

The distribution of particle sizes within the cloud regions strongly suggests the growth of acid droplets although there is some doubt as to whether any of the "acid rain" ever reaches the planet's surface. Lightning has been detected in Venus' atmosphere but its causes are not known. It may be produced by cumilonimbus-like storms but the probability that the electrical gradients are chemically induced seems more realistic.

Venus' lack of a magnetosphere revealed an ionosphere which compresses and expands with the varying strength of the solar wind. Whether the absence of a magnetic field results from the slow rotation rate or the lack of a liquid core remains an unanswered question.

The data from the Mariner and Pioneer probes answered many long standing questions about Venus. The data raised as many, if not more, questions than they answered. We shall have to content ourselves with the current information until NASA targets Venus for another probe.

References

Beatty, J.K., Pioneer's Venus: more than fire and brimstone, Sky & Telescope, 58, 13 ff., 1979.

Hoffman, J.H. et al., Composition and structure of the Venus atmosphere: results from Pioneer Venus, Science, 205, 49 ff., 1979.

Kerr, R.A., Venus: not simple or familiar, but interesting, Science, 207, 289 ff., 1980.

Kliore, A. et al., The polar ionosphere of Venus near the terminator from early Pioneer Venus Orbiter radio occultations, Science, 203, 765 ff., 1979.

Kliore, A.J., Initial observations of the night side ionosphere of Venus from Pioneer Venus Orbiter radio occultations, Science, 205, 99 ff., 1979.

Knollenberg, R. and Hunten, D., Clouds of Venus: particle size distribution measurements, Science, 203, 792 ff., 1979.

Knollenberg, R. and Hunten, D., Clouds of Venus: a preliminary assessment of microstructure, Science, 205, 70 ff., 1979.

Oyama, V. et al., Venus lower atmosphere composition: analysis by gas chromatography, _Science_, 203, 802 ff., 1979.

Pollack, J. and Black, D., Implications of the gas compositional measurements of Pioneer Venus for the origin of planetary atmospheres, _Science_, 205, 56 ff., 1979.

Pollack, J.B. et al., Nature of the ultraviolet absorber in the Venus clouds: inferences based on Pioneer Venus data, _Science_, 205, 76 ff., 1979.

Ragent, B. and Blamont, J., Preliminary results of the Pioneer Venus nephelometer experiment, _Science_, 203, 790 ff., 1979.

Russell, C.T. and Brace, L.H., Evidence for lightning on Venus, _Nature_, 279, 614 ff., 1979.

Sieff, A. et al., Structure of the atmosphere of Venus up to 110 kilometers: preliminary results from the four Pioneer Venus entry probes, _Science_, 203, 787 ff., 1979.

Seiff, A. et al., Thermal contrast in the atmosphere of Venus: initial appraisal from Pioneer Venus probe data, _Science_, 205, 46 ff., 1979.

Taylor, F.W. et al., Infrared remote sounding of the middle atmosphere of Venus from the Pioneer Orbiter, _Science_, 203, 779 ff., 1979.

Taylor, F.W. et al., Temperature, cloud structure and dynamics of the Venus middle atmosphere by infrared remote sensing from the Pioneer Orbiter, _Science_, 205, 65 ff., 1979.

Taylor, F.W. et al., Polar clearing in the Venus clouds observed from the Pioneer Orbiter, _Nature_, 279, 613 ff., 1979.

Taylor, H.A. et al., Ionosphere of Venus: first observations of the effects of dynamics on the day side ion composition, _Science_, 203, 755 ff., 1979.

Taylor, H.A. et al., Ionosphere of Venus: first observations of day/night variations of the ion composition, _Science_, 205, 96 ff., 1979.

Taylor, W.W.L. et al., Absorption of whistler mode waves in the ionosphere of Venus, _Science_, 205, 112 ff., 1979.

Tomasko, M.G. et al., Absorption of sunlight in the atmosphere of Venus, _Science_, 205, 80 ff., 1979.

Travis, L.D. et al., Orbiter cloud photopolarimeter investigation, _Science_, 203, 781 ff., 1979.

Travis, L.D. et al., Cloud images from the Pioneer Venus Orbiter, Science, 205, 74 ff., 1979.

Young, A. and Young, L., Venus, Scientific American, 233, 70 ff., 1975.

CHAPTER IV

MARS

Descriptive Statistics:

Distance from the Sun	1.5 A.U.
Radius	3,410 km
Period of Rotation	24 hours, 37 minutes
Mass	0.12 of earth
Density	3.96 g cm^{-3}

Introduction

Mars has intrigued earth based observers for centuries. It is visible with the naked eye as a ruddy-red colored dot amidst the other celestial objects. Telescopically, it has always been a difficult planet to observe. The poles are readily apparent but Mars' surface geology and atmospheric characteristics remained hidden from our view. The Mariner flyby and the Viking Mission, consisting of two orbiters and two landers, returned a data payload almost beyond the mission scientists' dreams. The process of comprehending Mars has begun but many unanswered questions remain. More often than not, the specific information sought answers one question and raises two new ones.

Atmospheric Composition

The primary constituents of the Martian atmosphere are carbon dioxide (95.3%), nitrogen (2.7%), and argon (1.6%). Lesser constituents, in order of decreasing amounts, are oxygen (0.13%), carbon monoxide (0.07%), and water vapor, amounts of which vary significantly planet-wide (0.03%). Trace constituents include neon (2.5 ppm - parts per million), krypton (0.3 ppm), xenon (0.08 ppm), and ozone (0.03 ppm).

The present day atmospheric composition is a function of processes operative in a planet's primordial atmosphere. Carbon dioxide dominates Mars' atmosphere and amounts to 13 g cm^{-2}. Studies relating carbon dioxide amounts to argon quantities indicate that

Mars may have actually outgassed 250 g cm^{-2} of carbon dioxide or approximately twenty times the present amount in the atmosphere. This difference implies that most of the carbon dioxide has been lost, although no one seems certain of where it went. Various theories suggest three possibilities. Mars may have lost its excess carbon dioxide directly to space. It is possible the carbon dioxide was outgassed and incorporated into the planet's rock materials via chemical weathering processes. Or, lastly, the missing carbon dioxide may have been physically adsorbed onto the soil particles. The data are not adequate to resolve the question of where the carbon dioxide did go at this time. The earth also outgassed large quantities of carbon dioxide but, unlike Mars, the bulk of it was incorporated into sedimentary, carbonaceous materials such as limestones, coal, etc. What remained was absorbed by the oceans. Approximately 0.033 percent remains in the earth's atmosphere. Yet, this seemingly small quantity of carbon dioxide (along with water vapor and ozone) is responsible for the absorption of infrared energy and the heating of the lower atmosphere.

The total amount of carbon dioxide in Mars' atmosphere undergoes strong seasonal variations associated with the formation of the polar caps. The carbon dioxide frost point is reached at temperatures of $\approx 148^{o}K$. When these temperatures are achieved as radiational cooling proceeds over the winter polar caps, carbon dioxide begins condensing out onto the surface. Estimates suggest that from one sixth to as much as one quarter of the atmospheric carbon dioxide condenses semiannually on the polar cap in the winter hemisphere. The actual growth of the polar caps and the effect of the condensing carbon dioxide on surface atmospheric pressure values will be covered later in this chapter.

Nitrogen is the second most common atmospheric constituent on Mars (2.7%) though it constitutes the bulk (78%) of the earth's atmosphere. The current ratio of $^{15}N/^{14}N$ in the Martian atmosphere is more than 1.6 to 1.7 times the ratio found on earth. Nitrogen gas (N_2) should lack the kinetic energy necessary to escape Mars' gravity, however, it could be accelerated to sufficiently high velocities by impact dissociation by electrons in the exophere. The lighter ^{14}N would selectively escape the planet's gravitational field leaving an enrichment of the heavier ^{15}N behind in the atmosphere.

Models based on the $^{15}N/^{14}N$ ratio suggest that nitrogen outgassing occurred on Mars beginning several billion years ago. Estimates suggest that Mars has outgassed between ten and thirty times the amount of nitrogen it now has. This quantity would have added more than 30 mb to the surface atmospheric pressure.

Argon, one of the noble gases, is present at the 1.6 percent level in the Martian atmosphere. It is present at significantly lower levels in the earth's atmosphere. Typically, the argon ratio between $^{40}Ar/^{36}Ar$ is approximately one hundred times greater on Mars than on the earth, however, this particular ratio must be used with caution. The decay product of ^{40}K (potassium-forty) is ^{40}Ar. Therefore, the quantity of ^{40}Ar measured in a planet's atmosphere may not represent the primordial inventory of argon the planet possessed but rather, may simply reveal higher potassium levels present in the planet's regolith.

Oxygen levels are extremely low on Mars (0.13%). There may be vast quantities of oxygen (in oxide form) chemically bound into the regolith but there is a notable paucity of it in the atmosphere. Atmospheric oxygen is generally associated with photosynthesis. The present Martian landscape appears devoid of living organisms. The absence of plants results in very little free atmospheric oxygen.

Carbon monoxide is a product of photodissociation of the abundant carbon dioxide on Mars. Neon, krypton, and xenon on Mars are found in the same relative proportions as they are found in the earth's atmosphere. Interestingly enough, ^{129}Xe (xenon) is much more abundant than either ^{131}Xe or ^{132}Xe. The ratios between isotopes of xenon are related to sedimentary processes on earth. Preferential adsorption of xenon in sedimentary materials has been observed. The preferential adsorption is more evident in those earth locations where shales have been metamorphosed into slate. There is a large body of circumstantial evidence for fluvial processes, at least historically, on Mars. There does not appear to be any evidence that would suggest even the historic remains of oceans or lakes. In the absence of sedimentary processes, the fluvial processes may account for the adsorption of heavier xenon isotopes (^{131}Xe, ^{132}Xe) and explain the abundance of lighter ^{129}Xe in the atmosphere.

The water vapor content of the atmosphere varies in quantity. It also varies latitudinally from the equator to the poles. The bulk of the water vapor is concentrated in the equatorial latitudes and in the northern hemisphere of Mars. The Martian atmosphere is, by earth's standards, quite dry. It would be more suitable to compare the atmospheric water vapor content on Mars and the earth at equivalent atmospheric pressure levels. In doing so, it becomes obvious that Mars' atmosphere is close to saturation levels planet-wide. A more detailed coverage of the amounts and the global distribution of water appears later in this chapter.

Very small amounts of ozone are present in the Martian atmosphere (0.03 ppm). This amount is approximately four times the amount of ozone above the earth's stratopause. Mars receives ≃43 percent of the ultraviolet energy received by the earth, however, this amount is sufficient to produce ozone (O_3). The ozone is produced by the same set of chemical reactions described in chapter one. Ozone has been detected during the late summer through winter period at high latitudes. Amounts decrease in the spring and ozone is undetectable in the summer. To date, ozone has not been detected in the equatorial regions at any time of the year. There is some speculation that ozone may be a very significant gas in the Martian atmosphere because it may affect the temperature regime of the high latitude regions. These subtle temperature changes may, in turn, control the rate of carbon dioxide deposition at the poles. The possibility exists that the greater ozone values in the northern hemisphere inhibit the polar deposition of carbon dioxide. The smaller amounts observed in the southern hemisphere appear to exert little or no effect on both the polar temperature regime and the carbon dioxide deposition rate suggesting that some critical level of ozone must occur before any effect on the carbon dioxide deposition rate occurs.

The compositon of Mars atmosphere raises a number of questions concerning its evolution. Since Mars and earth occupy adjacent regions of our solar system, it would be logical to suspect that they both began with the same inventory of volatile gases. The present day inventories are so different that it has been suggested that a catastrophic event altered Mars' atmosphere. If so, its proximity to earth suggests that earth should have experienced this event too. Suggestions that Mars has degassed less than the earth cannot be reconciled with the quantities of nitrogen almost certainly lost from its primordial atmosphere. The noble gas inventory leads to speculation that perhaps the volatile inventories on the earth and Mars were originally different. This, coupled with a lesser rate of degassing for Mars, can explain the differences and similarities between the two planets.

The atmospheric chemistry of Mars is based on two distinct cycles, both initiated by ultraviolet energy from the sun. The cycles begin with carbon dioxide and water vapor and are referred to as the oxygen cycle and the hydrogen cycle, respectively. In the oxygen cycle, carbon dioxide (CO_2) is dissociated by ultraviolet wavelengths into carbon monoxide (CO) and atomic oxygen (O). Numerous atomic oxygen form molecular oxygen (O_2) and triatomic oxygen (ozone) (O_3). Atomic oxygen and ozone are oxidants and may contribute to the oxidation of the Martian regolith. The hydrogen cycle begins with water vapor which is also dissociated by ultraviolet energy into atomic hydrogen (H) and the hydroxyl radical (OH). Reactions between available atomic hydrogen and available oxygen produce another powerful oxidant, hydrogen peroxide (H_2O_2). Hydrogen peroxide may also react to oxidize soil materials. In the end, carbon monoxide (CO) recombines with atomic oxygen (O) to produce carbon dioxide (CO_2) which explains why carbon dioxide remains the main constituent in the atmosphere. Some of the dissociated hydrogen and oxygen is lost to space. Since these gases originated from the dissociation of water, their loss represents a net water loss to the planet, i.e. dessication.

This dessication process may explain where the outgassed water from Mars' interior went. It may also help to explain why so many surface features on Mars appear to have been formed by fluvial processes and are currently preserved in an arid environment. We shall return to this idea shortly.

Albedo

The albedo of Mars varies considerably and irregularly. Historically, tonal changes were interpreted as "canals" or due to vegetative changes on the planet's surface. These myths are no longer viable as the Mariner and Viking probes revealed that changes in the Martian surface are related to aeolian, or wind induced, processes in combination with the deposition and/or sublimation of volatiles on the polar caps.

Aeolian surface changes on Mars are associated with either streaks or splotches (Fig. 4.1). The streaks may be bright or dark in color.

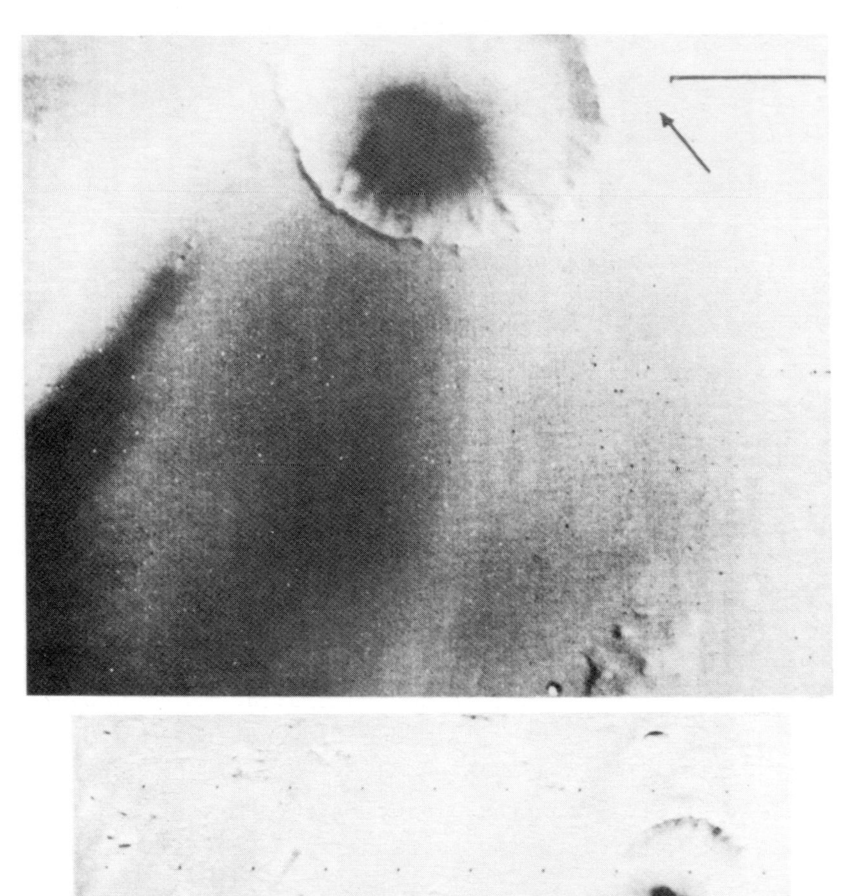

Fig. 4.1. Albedo changes within a dark streak. Upper image taken by Mariner 9 on June 26, 1972. Lower Viking image acquired on September 17, 1976. (Courtesy J. Veverka and P. Thomas, J. Geophys. Res., v 82, p. 4180, Sept. 30, 1977, copyrighted by the American Geophysical Union.)

The bright streaks generally appear after major dust storms lending support to the idea that the material in these streaks consists of dust storm deposition on the lee sides of topographic obstacles. The dark streaks probably represent areas where fine, brighter surface material has been transported away exposing darker basement rocks or subsoil. These dark streaks are relatively permanent from the time of their creation until their modification by the next global dust storm. The orientation of these streaks mirror the prevailing pattern of surface winds, particularly during the dust storm season associated with the period between the late southern hemisphere spring and the early southern hemisphere fall.

Dark patches on the floors of craters nearest the side where dark streaks begin are referred to as splotches. Splotches appear to be both erosional and depositional features. Some represent areas where the erosion of lighter colored surface material has exposed the darker basement rocks. Other splotches appear to be areas where darker colored material has been deposited. A few splotches coincide with transverse dune fields peripheral to the polar cap.

The wind speeds necessary to move and lift surface materials on Mars must be >50 m sec^{-1}. The amount of material eroded or deposited has been estimated as only several micrometers (μm) thick. Surface particles appear to be differentiated on the basis of color and size. Since the dust storms deposit brighter colored dust, it is probably safe to assume these bright dust particles are also the smallest in diameter. More importantly, the bright dust particles are probably the least dense of all soil particles and hence more easily carried aloft by the winds.

Vast dust clouds literally obscure the entire planet for extended periods of time ranging from a few weeks to several months. Albedo changes associated with global dust clouds are to be expected. Similarly, albedo changes in the polar regions will vary in response to the latitudinal extent of the polar caps. Excluding polar cap albedo changes, the majority of albedo variations on Mars are a function of the opacity of the atmosphere, i.e. how much dust has been raised, and aeolian changes are associated with strong surface winds.

Structure of the Atmosphere

The vertical temperature structure from the surface to an altitude of 200 km is presented in Fig. 4.2. There are few similarities

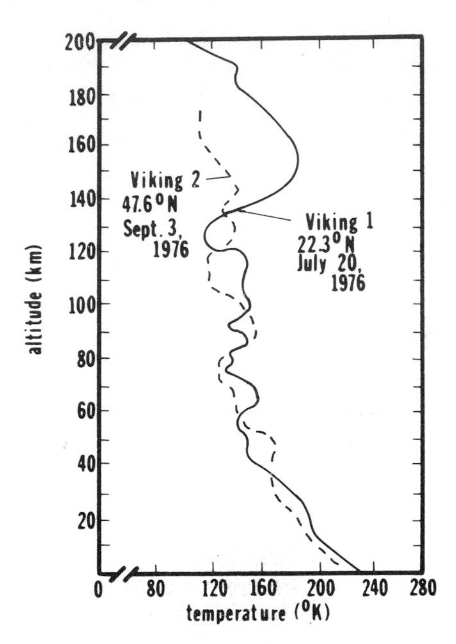

Fig. 4.2. Temperature profiles of the Martian atmosphere. (Courtesy A. Seiff and D. Kirk, J. Geophys. Res., v 82, p. 4373, Sept. 30, 1977, copyrighted by the American Geophysical Union.)

between this curve and the vertical temperature curve for the earth presented in chapter one (Fig. 1.1). Both temperature profiles reveal a temperature decrease from ground level to about 10 km, demonstrating that both atmospheres are warmed by reradiated thermal energy from the ground. Above 10 km, the Martian profile demonstrates a continuous decline to 40 km after which there appears to be a series of warmer and colder layers from 40-80 km. Actually, this portion of the curve is almost isothermal. The warmer and colder layers are a result of large amplitude waves attributed to the diurnal solar tides. These layers represent expansional cooling or compressional heating of the layers of gas and are associated with the daily passage of the sun. The increase in temperature with height in the earth's stratosphere and an associated ozone layer is completely lacking in the Martian profile. The upper portion (above 100 km) of the earth's temperature profile demonstrates the intense heating of a relatively few atoms through the absorption of ultraviolet energy. This increase in temperature associated with the earth's thermosphere is also absent in the vertical temperature profile of the Martian atmosphere.

The reported lapse rates in the Martian atmosphere vary, depending on the data source. Typically, lapse rates of between 1.6°-1.8°K km^{-1} in the lower 40 km have been observed. These rates are subadiabatic and appear to confirm that large scale convective mixing processes do not dominate the Martian atmosphere. This is not meant to imply that the atmosphere is not well mixed. Non-spherical dust particles with mean radii of 0.4 μm have been detected in the lower 30-40 km of the atmosphere, even on "clear" days. Their presence is evidence that the atmosphere is efficiently stirred by the winds. What it does mean is that convective processes probably play a more limited role on Mars than in the earth's atmosphere. The general state of the atmosphere from 0-40 km is subadiabatic, however, significant variations have been observed, particularly during dust storms. Significant temperature inversions appeared at 20-30 km during one of the global dust storms monitored by the Viking Landers in 1977. Lapse rates of -3.5°K km^{-1}, indicating a warming trend or inversion with increasing altitude, were reported. The dust increases the opacity of the atmosphere and limits the quantity of direct solar energy reaching the surface of the planet. Instead, the dust absorbs the incoming solar energy. The temperature of the dust and the dust cloud increases and reradiates the energy in the infrared wavelengths. The inversions at 20-30 km are a response to this process. The dust also radiates approximately one half of the energy it absorbs downward, toward the surface. This energy is ultimately absorbed by the surface since the Martian atmosphere is transparent to the penetration of infrared energy. The dust produces a net cooling in the vertical temperature profile between 0-30 km, but the amount of cooling is dependent on the quantity of dust raised, the area of the atmosphere affected and the duration of the event. Details concerning dust storm life cycles appear later in the chapter.

The vertical temperature profile reveals the existence of the Martian ionosphere near 130 km. A distinct F_1 layer with peak ion concentrations coincides with this altitude. The electron density has been reported as between 1 and 2 x 10^5 cm^{-3}. Oxygen ions (O_2^+, ≃90%) and carbon dioxide ions (CO_2^+, ≃10%) account for the bulk of the mass at these altitudes. Ion temperatures of $150^{o}K$ have been observed in the F_1 layer. Temperature values increase vertically to $210^{o}K$ at 175 km (in the exosphere) and $2,900^{o}K$ near 290 km. The dominant ion above 130 km is oxygen (O^+) with peak concentrations detected near 225 km. The altitudes at which these characteristics occur probably vary in relation to solar activity and Martian meteorological events. At the height of one dust storm, the electron peak was at 140 km but after the atmosphere cleared, the electron peak was found at 115 km. In general, there will always be some variations in the heights and temperatures of these layers.

Temperature variations were reported by the Viking Lander One (hereafter VL-1) and Viking Lander Two (hereafter VL-2). VL-1 is located at $22^{o}N$, $48^{o}W$ and VL-2 at $48^{o}N$, $226^{o}W$. The combination of lander and orbiter data revealed three major differences in the temperature characteristics at the two sites. The geographic distance between VL-1 and VL-2 accounted for the diurnal surface temperature differences. Obviously, it was generally warmer during the day and night at the VL-1 site located closer to the Martian equator. Seasonal and latitudinal variations accounted for the 3°-$10^{o}K$ temperature differences observed in the vertical temperature profiles below 35 km at each site. A third $25^{o}K$ difference in the vertical temperature curve above 120 km was attributed to atmospheric dynamics, particularly the atmospheric tides.

The VL-1 and VL-2 landers have observed temperatures for one full Martian year. Minima were near $190^{o}K$ and maxima near $242^{o}K$. Seasonal changes caused the VL-1 site maximum to reach values $>250^{o}K$ while at the VL-2 site, temperature values peaked at $205^{o}K$ and then slowly declined to minima of $170^{o}K$.

Minimum and maximum temperatures were also affected by the global dust storms. The VL-1 site experienced a sudden $10^{o}K$ decrease in maximum temperature and a $15^{o}K$ increase in the minimum values. A steady decline in both maximum and minimum temperatures occurred throughout the duration of the storm. Temperature extremes at the VL-2 site were barely affected by this particular dust storm.

A typical diurnal temperature profile is presented in Fig. 4.3A. The Martian curve is similar to a comparable desert location on earth in many ways. The diurnal minimum coincides with sunrise as the land surface radiated its energy out to space during the night. The

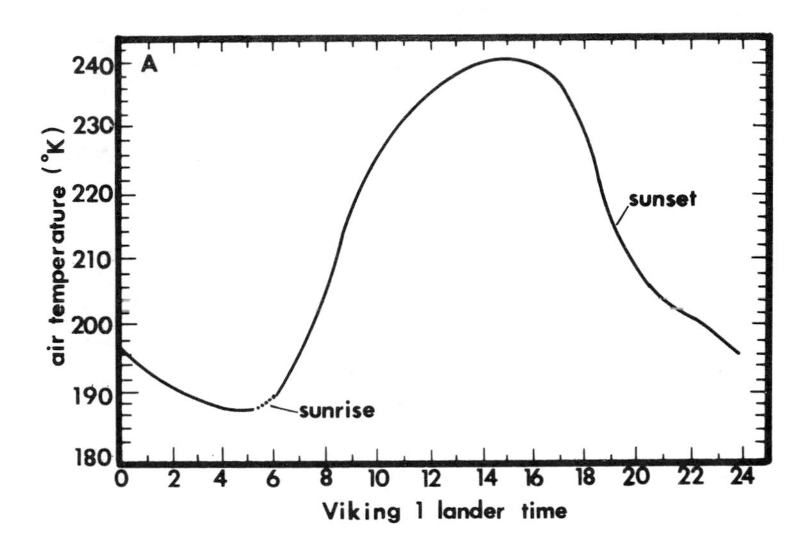

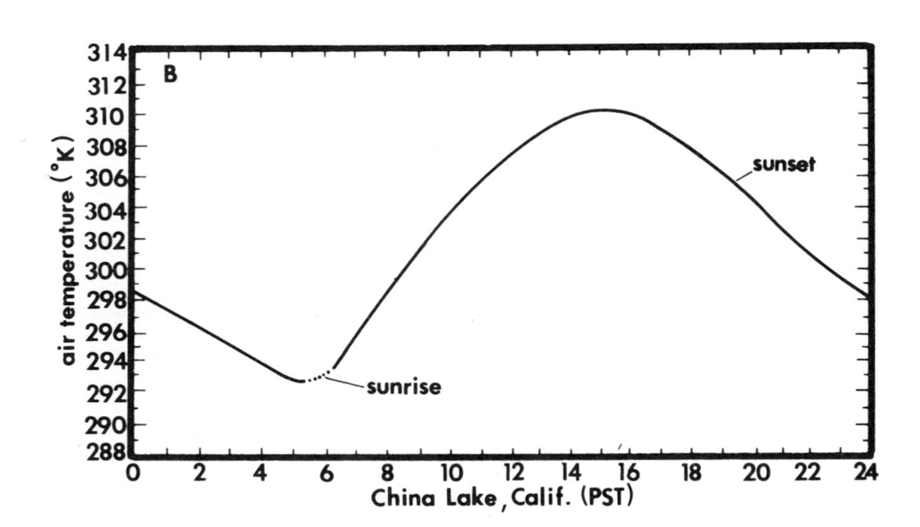

Fig. 4.3. Typical diurnal temperature profile at the Viking 1 lander site (A) compared to China Lake, California, a desert site (B). (Courtesy S.L. Hess et al., Science, v 194, p. 79, 1 Oct., 1976. Copyright 1976 by the American Association for the Advancement of Science.)

maximum occurred 2-3 hours after the peak solar energy receipt at noon. This location of the maximum temperature indicates a phase lag, i.e. the time it takes for the land to heat and reradiate infrared energy which then heats the atmosphere. This phase lag is similar to that observed on earth where maximum diurnal temperatures normally occur between 2-4 P.M. A comparison of the shape of the VL-1 curve with the profile from China Lake, California, a desert site, shows the trends are very similar (Fig. 4.3B). Differences exist in the actual values reported: 190°-240°K for VL-1 and 292°-310°K for China Lake. The diurnal temperature range at the VL-1 site is $\approx 50^{o}$K, whereas at China Lake, it is $\approx 18^{o}$K. The large diurnal range of temperatures on Mars is a result of the receipt of less total energy because of Mars' distance from the sun and a rarified atmosphere that permits rapid radiational cooling to proceed during the night time hours.

The temperatures on Mars attain lower minima and higher maxima than the values reported by the Viking Landers. Temperatures over the polar caps drop below the frost point of carbon dioxide (148°K) and achieve values $>250^{o}$K along the Martian equator. Thermal regimes are hemispherically asymmetrical on Mars. The planet receives ≈ 40 percent more solar energy at perihelion, when it is closest to the sun, than it receives when it is at aphelion or farthest away. The difference in the total energy receipt for the earth at perihelion and aphelion is only 3 percent. Summer in the Martian southern hemisphere coincides with perihelion. Summer in the southern hemisphere is warmer and more protracted than in the northern hemisphere. But, the present orientation between Mars and the sun represents only one example of the wide range of long period changes that may occur on Mars.

The present atmosphere of Mars varies within certain thermal limits yet nagging questions persist. Was Mars warmer and if so, did its atmosphere contain more water? Can climatic change explain the presence of large fluvial-like valleys? How do we explain the layered sequences in the Martian polar caps? Climatic variations on Mars are, and apparently have been, associated with rotational and orbital perturbations.

Several orbital and rotational variations have been discovered. Long period gravitational effects between Mars, the sun, and the other planets create a shift in the longitude of perihelion. The cycle repeats every 72,000 years. The wobble in the polar axis results in the precession of the equinoxes, i.e. they occur at different times of the year. This cycle has a period of 180,000 years. These two variations result in a combination cycle of 51,000 years. The orientation of the planet's axis and distance from the sun has a profound effect on the thermal characteristics of summer and winter (Fig. 4.4). In one position, Mars experiences a cool, northern summer at aphelion and a warm, northern winter at perihelion. Half a cycle later, the northern hemisphere experiences a frigid winter at aphelion and a hot, northern summer at the perihelion position.

The orbital eccentricity of Mars undergoes two cyclic variations of 95,000 and 2,000,000 years. Mars' orbital eccentricity (e) varies from nearly circular orbit (e = 0.004) to a highly elliptical orbit (e = 0.14). Its present eccentricity is 0.093. Thus, one hemisphere may receive as much as 1.5 times more solar energy at perihelion than at aphelion. The various hemispheres may experience excessively warm summers, or cool summers depending on the planet's distance from the sun at that season.

Variations in the obliquity of Mars' axial tilt may have also played an important role in the changing climate of Mars. The present obliquity is $\approx 25^{\circ}$, similar to the earth's 23.5° axial tilt. This can vary from 15° to 35° on Mars. These variations in axial tilt may be responsible for as much as a 30-40 percent reduction in the amount of solar energy the polar caps receive. Calculations suggest that the polar regions may experience continual above freezing day time temperatures for periods of 40 days or more when the axial tilt approaches 35°. Tilt variations of the magnitude described would dramatically alter the receipt of solar energy on Mars and the paleoclimate as well. Historically, the polar caps may have melted and reformed frequently. Prior to the formation of the Tharsis Ridge and associated volcanics, the axial tilt may have ranged from 21°-45°. Certainly, evidence now in existence suggests that a detailed thermal history of Mars needs to be compiled and that current atmospheric and climatic conditions be reexamined in the light of historic events of past climatic epochs. Clearly, the current temperature data from the Viking mission represents a detailed

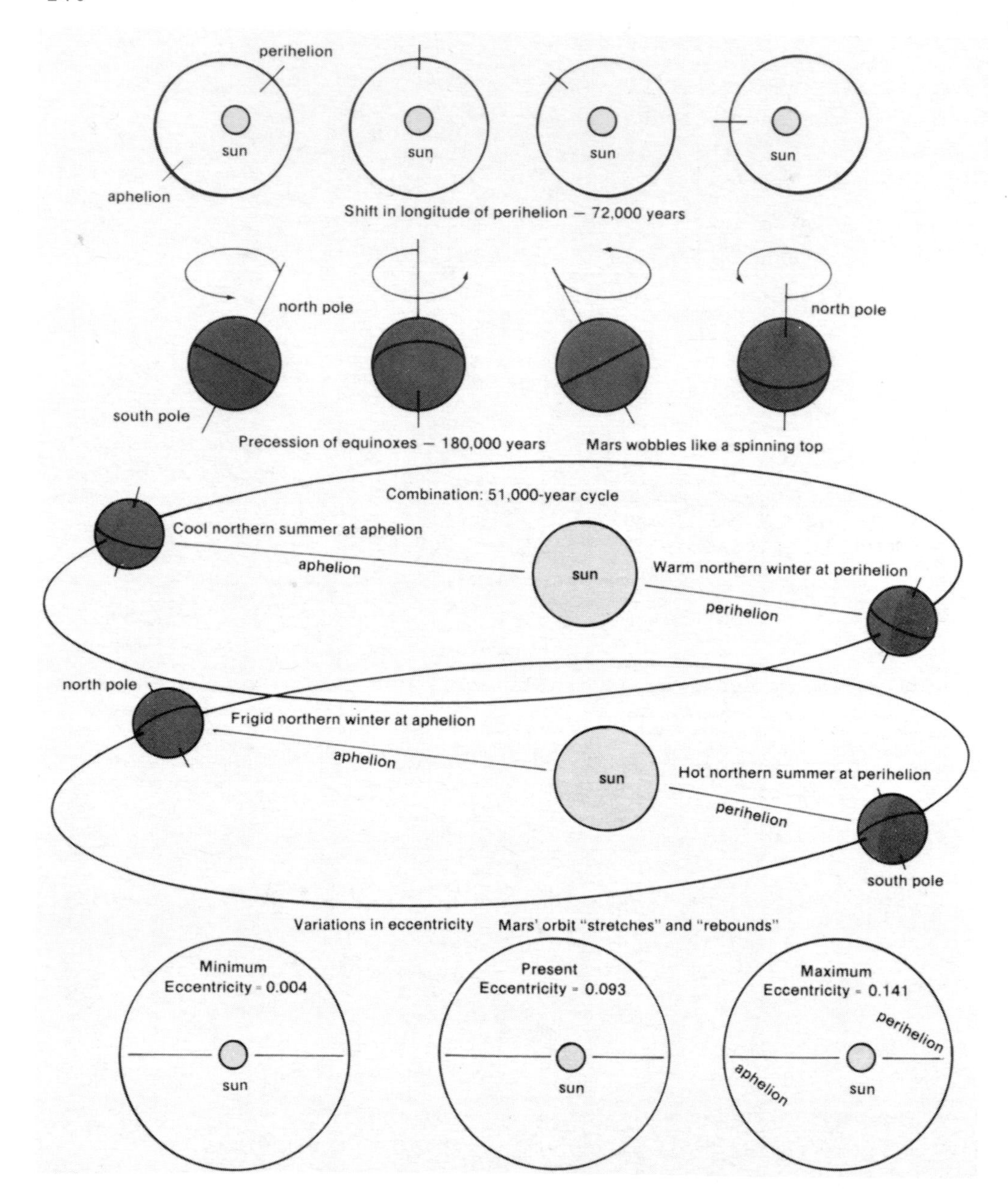

Fig. 4.4. Orbital effects on Martian climate. (Courtesy J. Gribbin, Astronomy, v 5, p. 22, Oct. 1977. Reproduced by permission of Astronomy magazine, copyright 1977 by Astro Media Corp.)

but limited view of the planet's energy relationships.

Atmospheric Pressure

The vertical pressure structure of the Martian atmosphere decreases linearly and ranges from a surface pressure of 7.0 mb to 10^{-9} mb at 200 km. Measurements from the VL-1 and VL-2 sites provided the first accurate surface atmospheric pressure data from Mars (Fig. 4.5). The pressure values from the VL-2 are consistently larger because the lander is at an elevation 0.96 km lower than the VL-1. Higher pressure values result as a greater thickness of the Martian atmosphere exerts a downward force on the pressure sensor. The minimum pressure observed at the VL-1 site was 6.9 mb and the maximum was 8.9 mb. Pressure values from the VL-2 are proportionately greater and reflect the site's lower elevation. Two minima are clearly visible in the pressure data. Sol 0 marks the touchdown of the VL-1. The first minimum occurred near sol 100 (Martian day 100) and coincided with late summer in the northern hemisphere and late winter in the southern hemisphere. The second, lesser minimum occurred one-half year later near sol 400. This coincided with late winter in the northern hemisphere and late summer in the southern hemisphere. These pressure variations were caused by the condensation of carbon dioxide onto the polar caps. Atmospheric pressure dropped as more carbon dioxide condensed and rose as carbon dioxide sublimated off the caps back into the atmosphere in the spring.

These annual pressure variations suggest that the equivalent of 23 cm layer of carbon dioxide sublimated from the south polar cap. This would correspond to 7.9 x 10^{15} km of carbon dioxide. If the Martian polar caps behave similar to the earth's continental ice sheets, the center of the cap should be thicker than the rim. The lesser pressure minimum associated with the northern polar cap indicates that it consists of 40 percent of the amount of carbon dioxide that sublimed of the southern cap.

Large diurnal pressure variations have also been observed at the VL-1 and VL-2 sites. These variations mark the passage of the atmospheric tides and wave phenomena associated with the strong daily heating of the relatively cloudless planet.

The pressure variations on Mars have been measured and cataloged. The largest variations are associated with the condensation and/or sublimation of carbon dioxide while the smaller diurnal variations

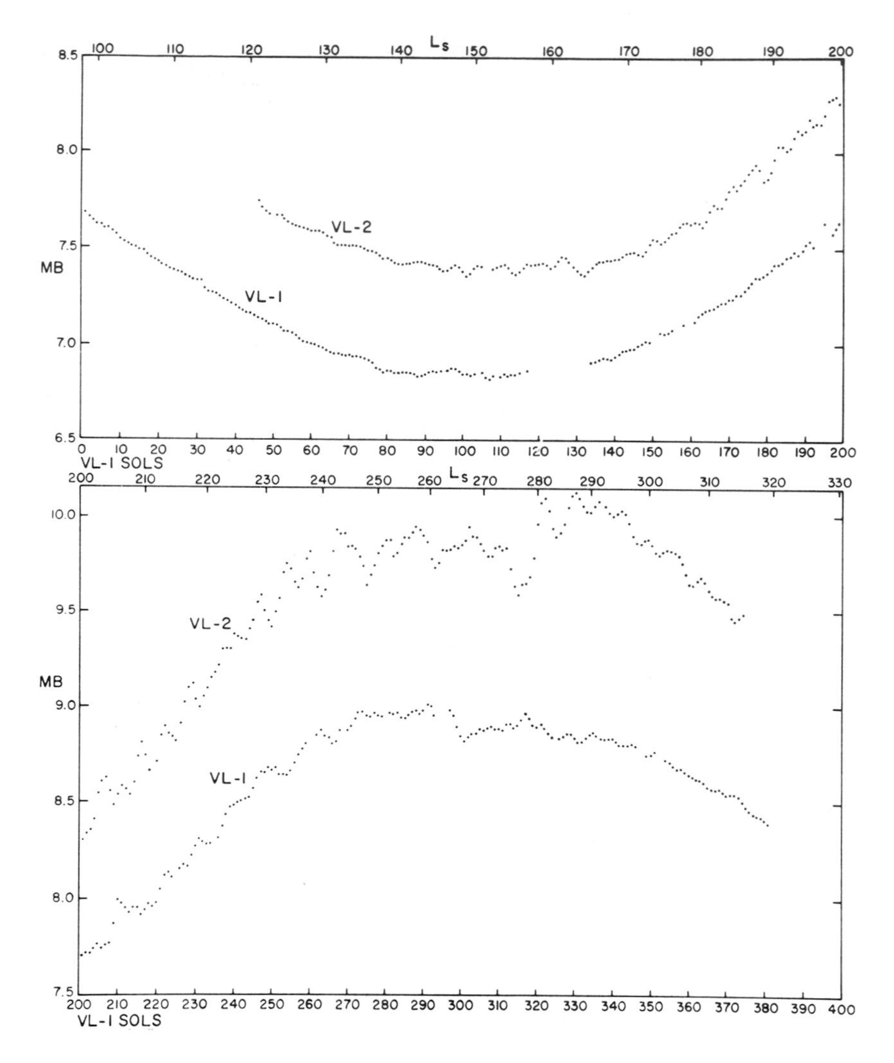

Fig. 4.5. Surface pressure variations at the VL-1 and VL-2 landing sites as a function of time. (Courtesy S. Hess et al., J. Geophys. Res., v 82, p. 2924, June 10, 1977. Copyrighted by the American Geophysical Union.)

are primarily associated with the daily atmospheric tides and wave phenomena. There are several possible long term effects on atmospheric pressure which remain possible, although they have not been observed. The idea that large amounts of carbon dioxide exist in a soil reservoir has led to speculation that cyclic, long period climatic changes might initiate gas exchanges between the soil and the atmosphere. Similarly, circumstantial evidence exists for the existence of permafrost on Mars. If a frozen, subsurface reservoir of water exists, a climatic warming might release more water vapor into the atmosphere. Either of these changes could exert a significant impact on the surface atmospheric pressure values. For the present, the atmosphere remains a thin envelope of gas around Mars.

Magnetosphere

The rarified nature of Mars' atmosphere as evidenced by the surface pressure data raised another question. Does Mars have a magnetosphere or does its atmosphere interact with the solar wind? Two different types of planetary/solar wind interactions have already been observed on earth and Venus. The earth's atmosphere is protected deep within the magnetosphere and it does not have any direct interaction with the solar wind. Instead, a bow shock is formed and the solar wind is heated, compressed, and deflected in the magnetosheath around the earth and its magnetotail. Venus lacks an intrinsic magnetic field. In its case, the solar wind interacts directly with its atmosphere. The Venusian ionosphere forms a bow shock as it compresses planetward. The solar wind is heated, compressed, and also deflected past Venus in the ionosheath. The magnetic pressure is sufficient to stand off the solar wind around the earth. Venus has a dense enough atmosphere so that its atmospheric pressure stands off the solar wind. In both cases, a pressure is exerted in the opposite direction of the solar wind flow sufficient to match the solar wind pressure and deflect it around the planet.

The Martian atmosphere does not possess enough atmospheric pressure to stand off the solar wind yet it seems to do so 40 km from the surface. Calculations of the pressures require an additional component to balance the solar wind pressure. An intrinsic magnetic field could provide the necessary pressure but direct evidence for one is lacking. An alternative would require an induced field based on the conductivity of the Martian ionosphere. The Russians claimed to have measured a bow shock but the acquisition techniques and the data are very confusing and cannot be considered reliable.

All evidence points to a weak planetary magnetic field, especially since it is known that the solar wind does not penetrate deep into Mars' atmosphere. Direct proof is lacking and will remain so until another probe is sent to Mars. It is probably reasonable to conclude that, in all likelihood, a weak, intrinsic, planetary field exists. The interaction of the solar wind with Mars' ionosphere and weak magnetic field would therefore represent a third class or an intermediate example (between Venus and the earth) of solar wind/ planetary interactions.

The existence of an intrinsic magnetic field is also considered strong evidence for a liquid core, a prerequisite for planetary magnetism. The massive Martian volcanoes testify to an historic period of lava flows supplied by magma at some depth. The existence of a liquid core would also indicate that tectonic and volcanic activity might not be extinct. However, more data are necessary before the planet's interior may be adequately modelled.

General Circulation/Winds

There are certain physical similarities between the earth and Mars. They are similar in radius and mean density. The earth rotates once every 24 hours whereas Mars rotates once every 24 1/2 hours. They both rotate from west to east and have comparable angles of axial tilt. Similar rotation periods and axial tilt result in similar seasons and a similar Coriolis effect. It has been very difficult to model the general atmospheric circulation of Mars, despite these similarities. The absence of large scale cloud bands whose speed and direction can be monitored has been one of the major obstacles. Computer models, using terrestrial analogs, predict a surface wind pattern that varies seasonally. The winter wind regime consists of mid-latitude westerlies, jet streams, and cyclonic storms. The other pattern consists of a summer equatorial regime characterized by gentle winds and very repetitive conditions based on thermal wind regimes.

Wind streaks have been used to determine the global wind patterns. These streaks, both bright and dark, are often tens of kilometers long and and a few kilometers wide. They represent direct evidence of strong surface winds because winds greater than 50 m sec^{-1} are necessary to move soil particles on Mars. These higher wind velocities which form the streaks occur during the period between late southern hemisphere spring and early southern fall when the southern latitudes are strongly heated by the sun.

The global distribution of wind streak directions is presented in Fig. 4.6. These arrows represent the prevailing strong wind directions and may therefore be regarded as a model of the surface circulation during the months coinciding with the southern summer. In the north, poleward of 40°, a general westerly wind pattern persists. The latitudes 0°-30°N experience northeast winds, however, upon reaching the equator, these winds become northerly. These winds flow poleward into the southern hemisphere and begin to deflect to the left because of the Coriolis effect. The prevailing winds between latitudes 0°-30°S are from the northwest. Winds off the southern pole are generally easterly but experience many variations. All surface winds are subject to local topographic control, particularly by the large volcanoes and the polar caps. The huge sand dune region surrounding the north polar ice cap consists of transverse and barchan dunes, indicative of a wind regime with more than one prevailing direction. Detailed images from the Viking Orbiter show a varied summer pattern around the north polar cap where winds may be from the pole or flow toward the pole, however, only on-pole winds occur during the winter and spring months. There is evidence for high wind speeds at certain times of the year, particularly winter. Wind speeds of >50 m sec^{-1} are necessary for saltation, (i.e. the physical bouncing of soil grains along the surface) on Mars. Winds capable of accumulating thick transverse dune fields observed near the north polar cap must be >75 m sec^{-1}.

The high altitude jet streams above the northern mid-latitude regions have velocities estimated at 100-120 m sec^{-1} at altitudes of 20-30 km. A reversal of these high altitude winds appears likely during the summer in the northern hemisphere. The upper level wind regimes are very speculative and are based on limited data. Their existence is certain but their detailed patterns necessarily await more data.

The summer equatorial wind regime is not totally restricted to the

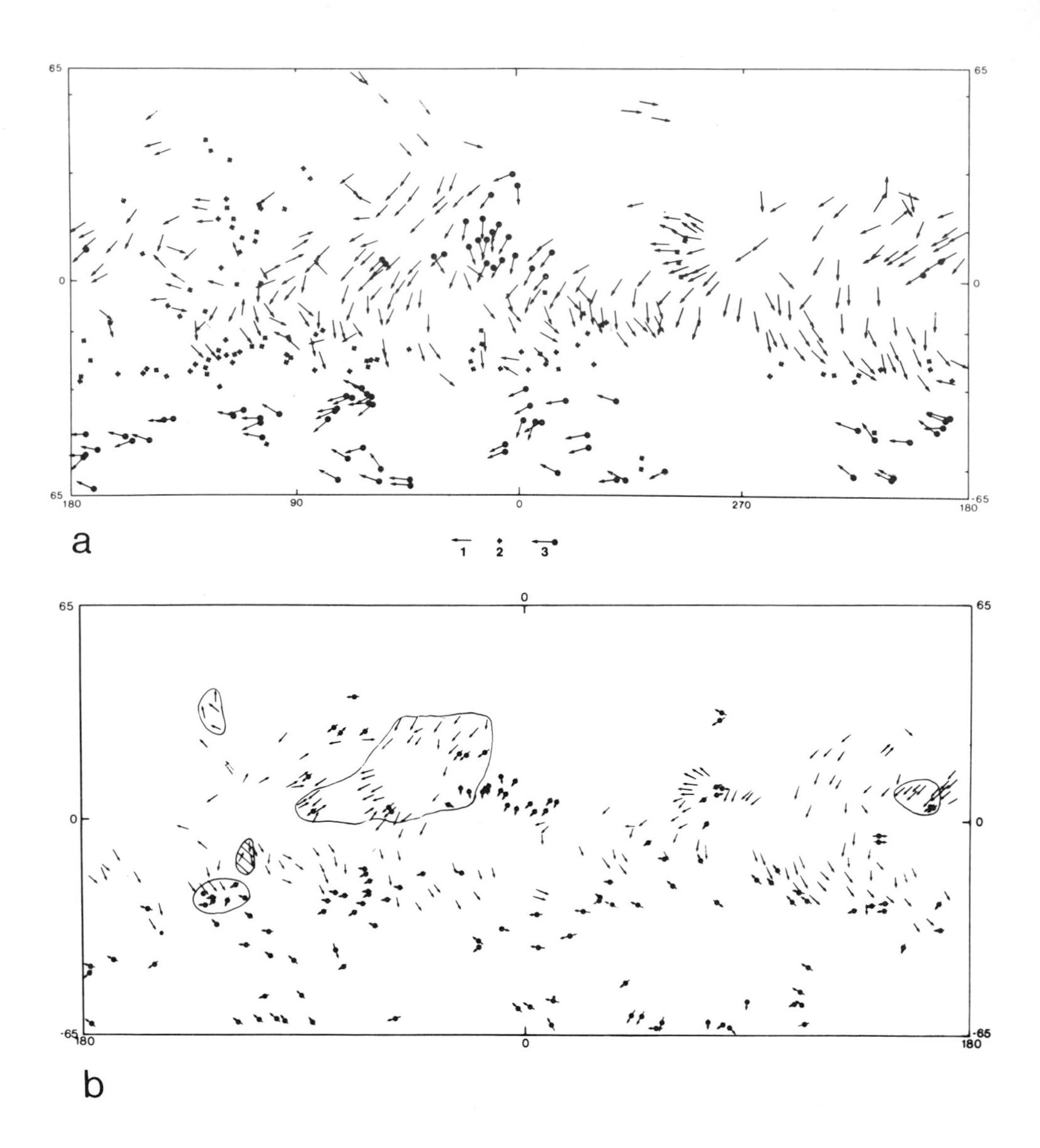

Fig. 4.6. Global distribution of wind streak directions from Viking data (A) and Mariner data (B). Bright streaks (1), dark erosional streaks (2), and dark splotch-associated steaks (3) are mapped. (Courtesy P. Thomas and J. Veverka, J. Geophys. Res., v 82, p. 8143, Dec. 30, 1979. Copyrighted by the American Geophysical Union.)

equatorial regions. This regime prevails poleward from the equator into the mid-latitudes of the summer hemisphere. The Viking Landers arrived on Mars in time to experience the northern hemisphere's summer. The VL-1 experienced a counterclockwise wind rotation every sol. The nocturnal winds were gentle, <2 m sec^{-1}. The winds were from the east at sunset and backed to southwest by midnight. These compass directions represent wind flow in a downhill direction at the VL-1 site. It is generally believed that the nocturnal winds represent drainage winds, i.e. the downslope movement of cooler air produced by radiational cooling. Wind speed increased after sunrise to velocities <7 m sec^{-1}. Later in the year, a trend to a stronger prevailing south-southwesterly flow was observed. The winds at the VL-2 site behaved differently than at the VL-1 site. Each sol, the wind rotated clockwise completing one revolution. Winds from all compass directions were recorded, whereas at the VL-1 site, no winds were observed from the north through east-northeast compass headings. Wind velocities were generally less than those observed at the VL-1 site. Maximum winds, which were from the southwest at VL-1, were from the southeast at VL-2. Throughout the summer months, these wind patterns repeated daily and confirmed that summer winds are primarily controlled by local constraints. Both the Vl-1 and VL-2 sites experienced gusty upslope winds during the afternoon and drainage winds during the period from midnight until dawn. This pattern is typical of the mesoscale wind flows in mountain and valley terrain on earth when the macroscale winds are gentle. Peak daytime surface gusts of 17 m sec^{-1} were reported from the VL-2 site. Several unusual wind and temperature events were observed by the landers, however, these will be discussed in the section on storms and fronts.

Water on Mars

The total amount of water outgassed from Mars has been estimated using the current loss rate of hydrogen from the exosphere. Since the hydrogen originated from the dissociation of water, these values permit calculations of the planet's lost inventory of water. Mars has apparently outgassed 10^4 g cm^{-2} of water throughout its existence. This is a vast quantity of water and is only one power less than the earth has outgassed (3×10^5 g cm^{-2}). Today, Mars is a desert planet, virtually devoid of water. Where did the water go? It is generally agreed that most of Mars' water was lost to space, however, there is enough evidence to suggest that some quantity of water may be locked up in the planet's regolith, well below the surface layer. The atmospheric water content on Mars has already been established as exceedingly small. Most of the atmosphere is near saturation levels not because of any surplus of water but because of the low minimum temperatures on the planet, particularly at night.

Moisture content varies latitudinally and seasonally with amounts generally ranging from 10-80 pr μm (i.e. precipitable micrometers). The maximum amount of water vapor coincides with the early summer of the northern hemisphere (Fig. 4.7). Over one half of Mars' water

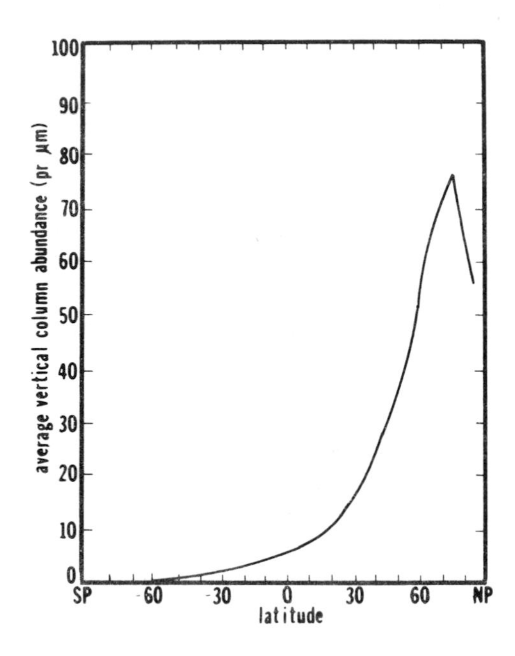

Fig. 4.7. Average distribution of global water vapor at 180°W. (Courtesy C.B. Farmer et al., Science, v 194, p. 1339, Dec. 17, 1976. Copyright 1976 by the American Association for the Advancement of Science.)

vapor is found poleward of 40°N while the latitudes poleward of 40°S are virtually devoid of water. The largest amount measured was 100 pr μm over the edge of the residual north polar cap also during the northern hemisphere's summer. The water vapor was not concentrated close to the surface, as had been expected, but was well mixed to altitudes of 10 km. This vertical vapor distribution is also evidence for turbulent and convective mixing on Mars. Water vapor amounts decreased where elevations changed abruptly but varied very little with gradual upslope changes. With the atmosphere close to saturation and always below the freezing point, phase changes always occur from vapor to solid and back. The present conditions eliminate the possibility of liquid water except under the most unusual circumstances.

The quantity of water vapor is dramatically affected by the global dust storms. During the summer months in the southern hemisphere, in 1977, the water vapor was evenly distributed between 60°N and 75°S. A peak of 25 pr μm was reported at 20°N. Onset of the first of two global dust storms in 1977 resulted in a 30 percent decrease in global water vapor. Water vapor inventories returned to pre-storm levels when the atmosphere cleared after the first storm. The second major dust storm reduced global water vapor content by 70 percent. When the storm had subsided, the vapor levels did not rebound uniformly to pre-storm levels. Over one third of the water vapor content was permanently removed from the atmosphere (Fig. 4.8). The northern hemisphere returned to pre-storm values but the southern hemisphere did not. Apparently, the water vapor that was removed from the southern hemisphere precipitated out as ice, possibly on dust particles, onto the north polar cap. This south to north transport of water vapor has been observed, however, there are no observations to suggest a return movement of water vapor from the northern to southern hemisphere.

The hemispheric sources of water vapor appear to be different. In the northern hemisphere polar regions, the water vapor content of the atmosphere is able to keep pace with the rising spring temperatures. This process occurs even though the quantities exceed the amounts of vapor elsewhere on Mars. This pattern suggests a supply of water somewhere in the northern hemisphere and provides indirect, but nevertheless convincing, evidence for a water reservoir somewhere in the regolith. The reservoir might consist of a layer of permanently frozen ground, i.e. permafrost. In the southern hemisphere, water vapor amounts increase in the springtime until the amounts equal quantities present over the equator. The amount of water vapor then remains consistent with equatorial values, even though the southern atmosphere continues to warm. This pattern

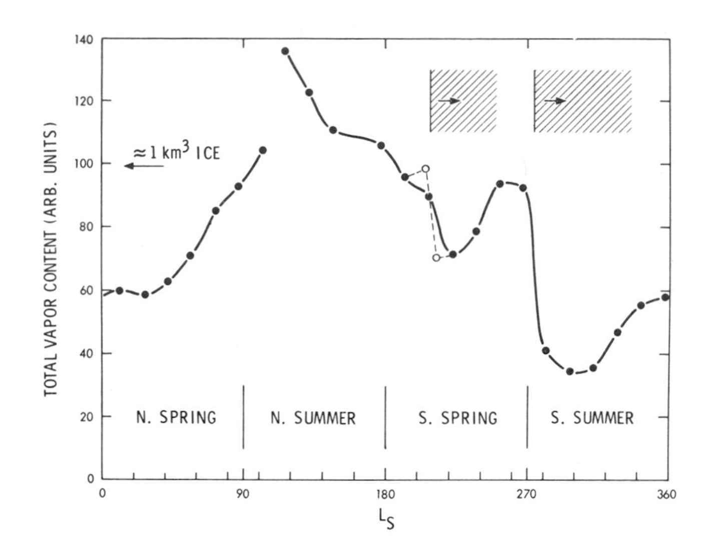

Fig. 4.8. Season variation of the global water vapor content. The occurrence and approximate durations of the two planetwide dust storms are indicated at the top of the figure. (Courtesy C.B. Farmer and P.E. Doms, J. Geophys. Res., v 84, p. 2886, June 10, 1979. Copyrighted by the American Geophysical Union.)

strongly suggests that the vapor supply for the southern hemisphere is a result of a poleward transport of moist air from the equatorial regions.

The behavior of water vapor amounts in the northern hemisphere clearly suggests the presence of a water ice reservoir, possibly a permafrost layer, at depths of 10 cm to 1 m poleward of 40°N. Several different types of surface features provide indirect support for a water ice reservoir. These features may be classified under the heading of polygonal ground, volcanic ejecta morphology, mass wasting, and thermokarst features.

Polygonal ground is a phenomenon associated with the land surface of a permafrost region. The surface appears cracked in a polygonal pattern when viewed from above. Much of the tundra landscape on earth exhibits this pattern of cracks and. in some locations, polygonal stone rings. The cracks form from the expansion and contraction of clays associated with repeated freeze/thaw cycles. The stones (on earth) are first forced to the surface and then, if any perceptible slope exists, moved downhill by the soil movement associated with the freeze/thaw cycles. Viking orbiter images show patterned ground patterns located on the northern plains of Mars. This region consists of well-defined polygonal cracks formed in level terrain. Many of the polygons are much larger than those found on earth, reaching 20 km in diameter. This pattern strongly suggests that these cracks are permafrost contraction cracks.

Studies of the craters produced by the ejecta from recent meteoric impacts clearly show that these craters are distinctly different from similarly formed craters on Mercury and the earth's moon. The ejecta craters' shape might be explained if ground ice (permafrost) was present in the Martian soil.

Mass wasting is the downslope movement of material by gravity. Landslides are a form of mass wasting readily apparent on the Viking photographs of Mars. One large landslide occurred on the southern wall of Gangis Chasma located at $9^{o}S$, $44^{o}W$. The slide moved 100 billion m^3 of material 60 km downslope at speeds estimated at ≈100 km hr^{-1}. It is difficult to comprehend a slide of this magnitude in dry material. It has been suggested that the materials contained ground ice which melted. This would certainly explain the more fluvial characteristics of this flow. The heat for the melting could have come from the sun or an internal heat reservoir, perhaps associated with subsurface reservoirs of warmer rocks or volcanic activity.

Thermokarst features on Mars are represented by a number of collapsed features that resemble the sinkholes associated with karst topography on earth. Limestones on earth are dissolved by groundwater producing a host of collapsed features termed karst topography. Sinkholes and caves are among the most common karst features on earth. The existence of a subterranean heat source could melt the permafrost and produce what are referred to as thermokarst or collapsed features on Mars. The behavior of atmospheric water vapor, the existence of polygonal cracks, landslides, and what appear to be thermokarst features present a convincing, if nonetheless circumstantial, argument for the presence of permafrost in the northern hemisphere of Mars.

Polar Caps

The asymmetrical water vapor distribution in combination with seasonal temperature and pressure changes control the growth and decay of the Martian polar caps. Mars reaches perihelion during the southern hemisphere's summer. The northern hemisphere's fall and winter seasons are not only warmer but also 75 days shorter due to the nature of Mars' orbital ellipse. These seasonal variations affect the geographic size of the polar caps as well as the quantity of materials condensed out of the atmosphere and deposited onto the polar caps. The southern polar cap is larger in response to the extended winter season which coincides with aphelion. The southern cap may have an angular extent as large as 100°. At its maximum, the southern cap could, if symmetrically developed, cover the southern hemisphere from 90°-40°S (pole to mid-latitudes). Its minimum width may range from 0°-5° in angular extent. The northern cap may achieve a maximum angular extent of 53°-80°. The lower limit for its minimum extent is 6°.

The north polar winter cap consists of condensed carbon dioxide from the atmosphere. This, as previously stated, results in a surface atmospheric pressure drop planet-wide as the atmosphere loses mass. Temperatures over the north cap must decrease to 148°K to reach the frost point of carbon dioxide. Temperatures actually remain below 148°K with minimum temperatures reaching 133°K. In the spring, air temperatures moderate and at temperatures above 148°K, the carbon dioxide sublimates back into the atmosphere. The residual cap of water ice also begins to sublimate and its size decreases as summer progresses. It was originally thought that the residual or permanent polar cap might be carbon dioxide or possibly a carbon dioxide-water clathrate ($CO_2 \cdot 6H_2O$). As previously stated, carbon dioxide cannot exist as a solid on Mars unless the temperature is below 148°K. The summer temperatures, therefore, ruled out the possibility of a carbon dioxide residual cap. The concept of a clathrate or combination of solid carbon dioxide and water ice where the carbon dioxide is inside the ice lattice in a cage-like structure proved interesting until the Viking Orbiter reported summer temperatures of 205°K over the residual north polar cap. The clathrate, if it ever existed, could not last at temperatures $>155^{o}$K. The 205°K temperatures and the large water vapor readings from the atmosphere over the northern cap proved that the residual cap was water ice. A number of estimates have been offered for the thickness of the residual cap. Generally, the thickness of the residual cap is thought to range from as little as 1 m to as thick as 1 km.

The albedo of the residual cap was measured at 0.43, a very low reading for an ice cap. Viking images revealed that the water ice also included large quantities of dirt and that the edge of the cap appeared to consist of thin sheets of horizontal layers. The layers appeared thin (5-50 m) and extended for hundreds of kilometers. The global dust storms were identified as the source of the dirt. The dust storms originate in the southern summer. If they become global, dust surges from the summer southern hemisphere into the northern winter hemisphere. The significant decreases in water vapor in the northern hemisphere in 1977 suggest that the dust grains act as condensation nuclei for water ice. The increased mass as the ice coats the dust grains causes the particles to precipitate out over the polar ice cap resulting in dirty ice layers with a low albedo. These "dirty" layers in the northern residual polar cap may provide an historic record of the global dust storms and climatic change on Mars, just as the ash layers in glaciers on earth record the evidence of episodic volcanic eruptions.

The longer, colder winters in the southern hemisphere result in a large polar cap and a decrease in surface atmospheric pressure, greater than that decrease produced by the north polar cap, as the carbon dioxide condenses out. The mass of carbon dioxide condensing from the atmosphere has been estimated at 7.9×10^{12} metric tons, or enough to cover the polar cap with a carbon dioxide layer $\approx$23 cm. The summer residual cap is not water ice but carbon dioxide. The cap is small and has a low albedo which should accelerate the sublimation of the carbon dioxide off the cap and into the atmosphere. The southern cap is preserved by the atmospheric dust which absorbs more of the sun's energy at higher altitudes above the cap. The net result is a cooler, more stable, lower atmosphere over the cap's surface which, in combination with the dust aloft, protects the residual summer cap.

Meteorological Phenomena

Dust Storms

Martian dust storms may be divided into two categories, local and global. Many smaller local dust clouds have been observed by the

Viking Orbiter. Several local storms will occasionally combine to inject large quantities of dust aloft. A global dust storm is formed when several local storms coalesce. Global dust storms are periodic events on Mars. Five global dust storms occurred on Mars between 1970 and 1980. They occurred in 1971, 1973, 1977 (2 storms), and 1979.

Dust storms originate in the southern hemisphere summer when Mars is at perihelion (Fig. 4.9). The southern hemisphere is strongly

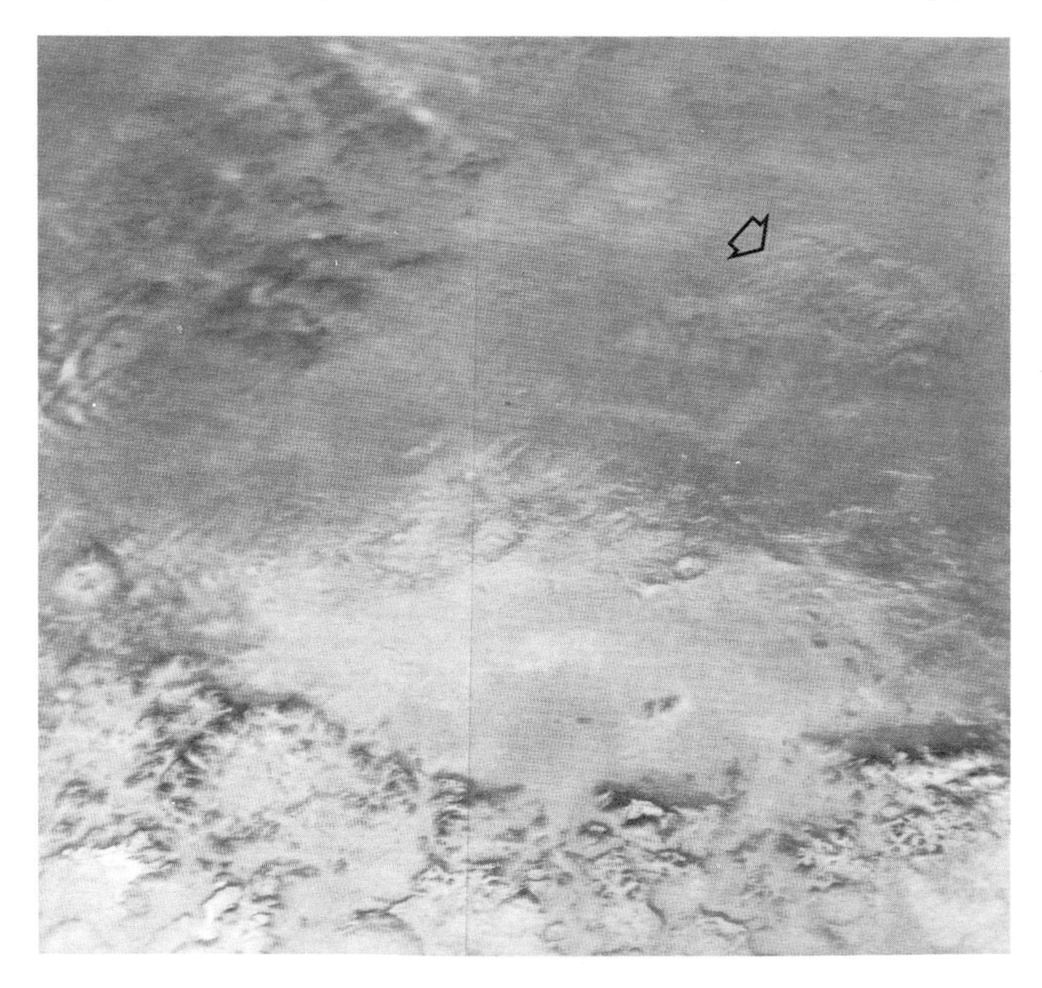

Fig. 4.9. A large dust storm in the Argyre Basin. The dust clouds are 4-6 km high and cover over 40,000 km^2. (Courtesy Jet Propulsion Laboratory/NASA.)

heated by the increased insolation and the atmosphere becomes less stable. Carbon dioxide is rapidly sublimed off the south polar cap. Strong winds are generated adjacent to the cap in response to the sharp thermal gradients between the cold cap and the surrounding land areas. Wind speeds >75 m sec^{-1} lift the dust into the atmosphere creating local dust storms. Global storms begin when several local storms lift large quantities of dust to altitudes of 30 km. The dust clouds begin to absorb the incoming solar energy directly and atmospheric temperatures increase aloft. The thermal gradients aloft increase wind speed and at some level of atmospheric opacity, the dust producing area grows. The dust is then transported rapidly toward the northern hemisphere. The total buildup and obscuration of the planet occurs in a period of three weeks.

It takes many weeks for the dust to settle out of the atmosphere, although the Martian atmosphere is never completely free of dust. The major mechanism for dust removal appears to be associated with condensation of water ice and carbon dioxide on the dust grains and their precipitation onto the growing north polar cap. At the same time, the global storm has its own self-limiting mechanism. The absorption of solar energy at 20-30 km in the dust clouds decreases insolation at ground level. The surface atmosphere slowly cools and stability increases. Wind velocities decrease below the critical wind speed of 75 m sec^{-1} and the source of dust is cut off. The atmosphere may slowly clear or, as clearing occurs, renewed surface absorption of solar energy increases surface temperatures. The atmosphere becomes increasingly unstable, wind velocities increase, and another global dust storm may begin. The second global dust storm of 1977 began in this manner.

The second global dust storm (1977) exerted a strong influence on the environmental conditions of the northern hemisphere. The VL-1 recorded increased wind velocities and a significant pressure increase. Wind velocities averaged 18 m sec^{-1} with gusts to >25 m sec^{-1}. The Viking Orbiter observed a 50°K increase in temperatures over the northern pole. At 72°N, polar night temperatures increased by more than 80°K. This warming aloft temporarily halted the growth of the north polar cap. Carbon dioxide may have actually begun to sublimate off the polar cap as evidenced by the pressure increase reported by the VL-1.

The life cycle of Martian dust storms reveals a series of interactions with many other meteorological phenomena. There is little doubt that dust storms can alter the energy transport characteristics of the atmosphere up to 30 km. They are also capable of transporting warmer air into the north polar regions and eventually contributing to the growth of the north polar cap as dust nuclei act as condensation nuclei for carbon dioxide and water ice which then precipitate onto the polar cap.

Clouds/Condensates

Condensation phenomena may take the form of fogs, frost, or clouds on Mars. Water ice ground fogs were suspected on Mars some time before their actual detection. Most of the atmosphere is near saturation, particularly during the night time hours. The Viking Orbiter took the photographs in Fig. 4.10A and Fig. 4.10B on July 24, 1976. The scene in Fig. 4.10A was photographed from an altitude of 12,400 km shortly after sunrise. Thirty minutes later, the orbiter rephotographed the same area from an altitude of 9,800 km (Fig. 4.10B). The arrows point to the fogs that developed shortly after sunrise on the valley bottoms and crater floors. The warming of the sun had apparently driven off some water from the soil. Upon contact with the colder air above the surface, the water recondensed in the colder air as a fog. This fog is similar to the phenomenon known as arctic sea smoke on the earth. Water will evaporate off a reservoir or sea surface, particularly in the fall, only to recondense upon contact with the colder air above as a fog. Estimates suggest that the water content of the Martian fogs in Fig. 4.10B represent a layer of water 1 μm thick. The fogs are probably 0.4 km thick. These photographs provide conclusive proof that fogs occur on Mars.

Two types of frost have been observed on Mars. Water ice frost and carbon dioxide frost. The carbon dioxide frost is less common than the water ice frost because temperatures must drop to 148°K or less for carbon dioxide frost formation. Temperatures this cold normally occur, if they are going to occur, a few hours before sunrise. The frost disappears shortly after sunrise. This type of frost has been observed within 15° of the equator in the Memnonia region. Carbon dioxide frost condensed out on the floor of craters less than 4 km in elevation. The only possible explanation for such low temperatures (148°K) in the equatorial region must relate to the

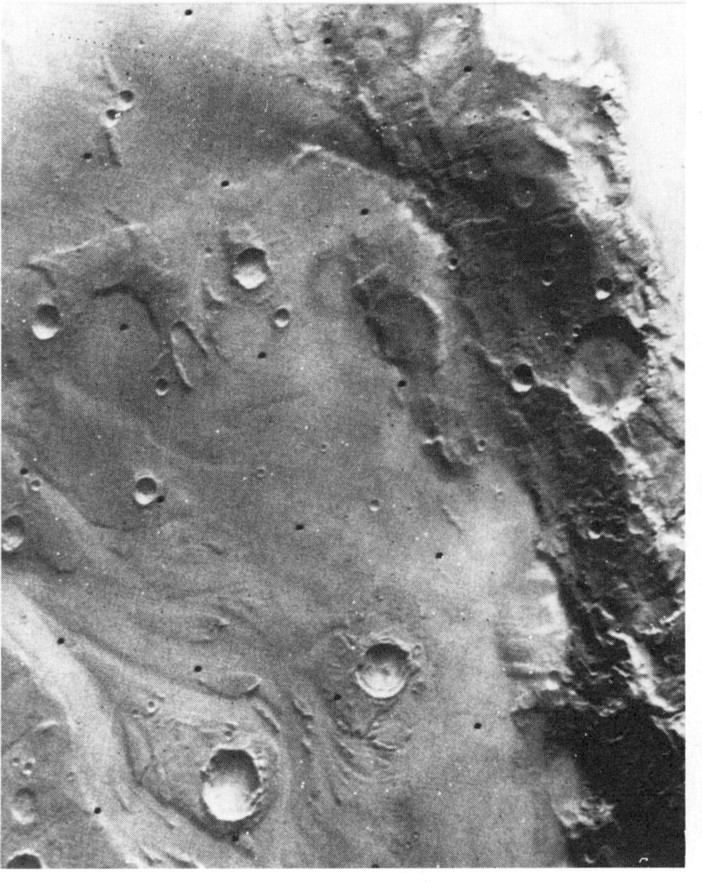

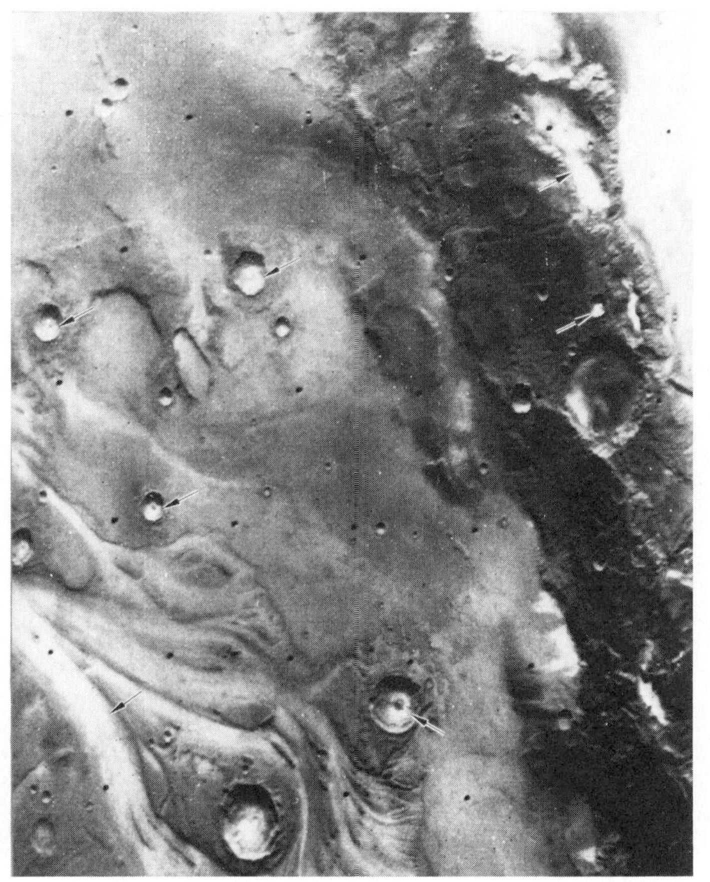

Fig. 4.10. Fog on Mars. The photograph on the left was taken at dawn. The photograph on the right was taken 30 minutes later. Arrows point to fog locations. (Courtesy Jet Propulsion Laboratory/NASA.)

materials present in the crater bottoms. This material apparently has a very low thermal inertia, i.e. it absorbs energy quickly and loses heat energy equally as fast. This rapid radiational cooling in a closed area (the crater) allows the temperature to plummet to the CO_2 frost temperature.

The VL-1 and VL-2 sensors detected water frost formation on many occasions, usually beginning from 1-3 hours after midnight at temperatures between 191°-196°K. A temporary halt to the decrease in nocturnal temperature was attributed to the frost cooling. A spectacular photograph of the frost coated Argyre Basin in the southern hemisphere is present in Fig. 4.11.

The VL-2 observed the formation of a thin water frost clathrate combination which lasted for more than 200 sols. The condensate settled out of the atmosphere on dust particles because when the frost coating began to disappear, a bright, but very thin, dust cover became apparent. The quantity of water in this frost probably amounted to less than several μm. When the condensate began to disappear, the southern facing areas were the first to clear. The areas in the rock shadows were the last to clear.

Clouds are the most common condensate forms on Mars. Wave-clouds typically form on the lee side of craters. The ship's wake pattern (Fig. 4.12) produced by the 100 km diameter crater Milankovic (53°N, 148°W) stretches downwind over 800 km. The waves are produced as the air is forced over the topographic obstacle, cooling adiabatically as it rises. On the downwind side, the air does not immediately return to a laminar flow pattern but instead oscillates up and down similar to a sine wave. The clouds form on the rise of the wave oscillation and reevaporate as the air subsides on the downward slope. Plume or scarf clouds have similar origins (Fig. 4.11). In Fig. 4.11, the air has been forced to rise over Ascreaus Mons, a volcano, and has condensed into a long plume-shaped cloud stretching hundreds of kilometers downwind. This type of cloud is often observed on earth during the winter months, trailing downwind from high mountains.

High altitude clouds often form at night in the mid-latitudes of the winter hemisphere where temperatures are $\approx 120^{o}$K. These clouds are thick enough to be easily detected but disappear within one hour after sunrise. Water ice clouds have also been photographed by the

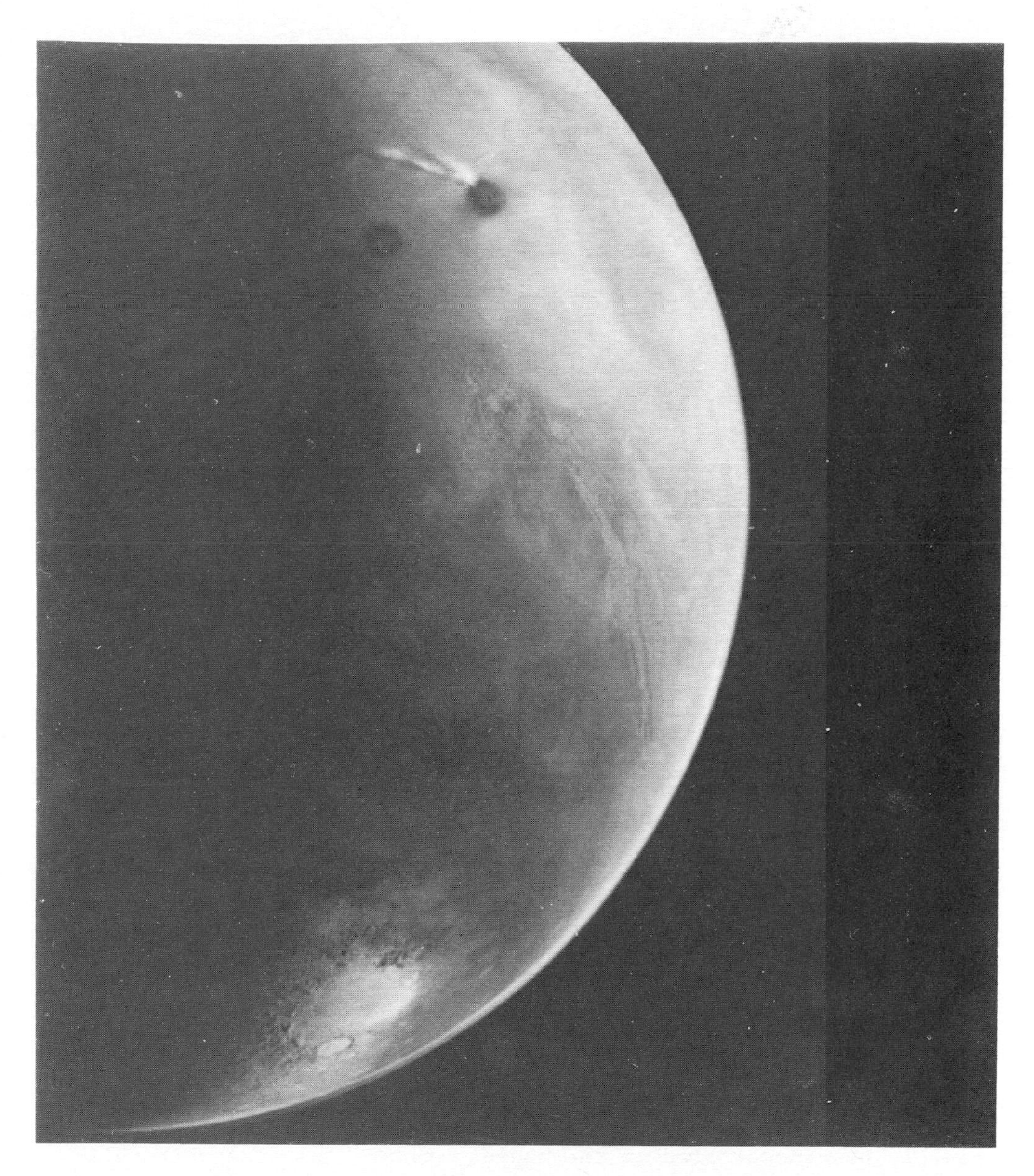

Fig. 4.11. Frost filled Argyre Basin just before dawn. Note water-ice plume cloud on the western flank of Ascreaus Mons. (Courtesy Jet Propulsion Laboratory/NASA.)

Fig. 4.12. Lee wake cloud system produced by the crater Milankovic (100 km in diameter). (Courtesy Jet Propulsion Laboratory NASA.)

Viking Orbiters at lower altitudes around the Noctis Labyrinthus (Fig. 4.13). The clouds in Fig. 4.13 are the white, bright areas located in the canyons and above the rust colored plateau. The clouds appear to be concentrated in the canyon areas. This suggests that water, which condensed in the shaded canyon areas during the previous afternoon, vaporized as the sun's rays warmed those regions the following morning.

The most significant and persistent of all clouds on Mars is the polar hood cloud. The polar hood is a condensation phenomenon that

Fig. 4.13. Water ice clouds in the Noctis Labyrinthus region taken by Viking 1 Orbiter. The area covered is ≈10,000 km^2 centered near 9°S, 95°W. (Courtesy Jet Propulsion Laboratory/NASA.)

occurs over the polar caps primarily during the winter months. In the northern polar regions, the hood extends from the pole to the edge of the polar cap which is usually near Mars' arctic circle. It disappears rapidly after the spring equinox. Dust storms modify the northern polar hood occasionally, causing it to disappear temporarily. The dust particles may act as condensation nuclei and precipitate the water ice and carbon dioxide from the polar hood out onto the polar cap.

A polar hood cloud also develops over the south polar cap late in the winter. The south polar hood is thinner and does not persist as long as the northern hood. It is possible that the lack of water vapor in the southern hemisphere may have some role in the limited development of the southern polar hood.

The VL-2 experienced several events which represented the southward advance of the Martian polar front. The front coincided with the edge of the hood cloud. This sequence of events included a rapid increase in atmospheric opacity so that the sun was not visible, a sharply defined leading edge, a retardation of the diurnal temperature increase, a pressure increase, and a wind shift from southerly to northerly. Clearly, a cold front passed the VL-2 site. The frontal zone was estimated to be 6 km wide with wind gusts ranging from 4-14 m sec^{-1}. The VL-2 has since recorded many frontal passages. These passages and variations in pressure values led scientists to suspect that cyclones and anticyclones were passing somewhere between the VL-2 and the polar cap. The Viking Orbiter photographed a cyclonic storm in the northern hemisphere quite similar to the extratropical cyclones experienced in the northern mid-latitudes on earth (Fig. 4.14). It eventually photographed four separate storms. The cyclonic storms were large and exhibited spiral clouds and a counterclockwise rotation. They were well developed and occurred only in early summer along the polar cap boundaries where relatively calm conditions persisted at that time of year. The clouds were composed of water ice and reached altitudes of 6-7 km. The discovery of cyclonic storms on Mars was perhaps, one of the most significant meteorological discoveries to date. Hopefully, modelling studies comparing earth cyclones with Martian cyclones will provide additional insights into the meteorology of Mars.

Fig. 4.14. An extratropical cyclone observed by the Viking Orbiter 1 from 29,000 km. Storm is at 65°N. (Courtesy Jet Propulsion Laboratory/NASA.)

Life On Mars

It is impossible to study Mars and not raise the question "Is there life on Mars?" Any search for life must necessarily center on identifying microorganisms in the soil. Certainly, the environmental parameters discussed in the preceeding sections of this chapter present a bleak picture for any "terrestrial-like" life forms. The atmosphere is low in oxygen and the atmospheric pressures and temperatures only permit what little water there is to exist in either the solid or vapor phase. The absence of organic materials in the soil samples causes one to wonder if any carbon-based organisms would stand a chance on Mars. But, Martian environments have not always been so harsh. Photographs of the landscape suggest that fluvial processes cut canyons and tributaries in past epochs. The existence of more water increases the possibility that life forms may have evolved on Mars. The current dessication is proof that Mars has somehow lost much of its water supply, however, the rate at which it lost the water is significant. A slow loss might have permitted microorganisms to adapt to the harsher environment whereas a rapid loss would not. The life experiments aboard the VL-1 and VL-2 were designed to search for microorganisms. They were designed with the assumption that microorganisms will release gas as they metabolize food. Typical gaseous by-products include carbon dioxide, nitrogen, hydrogen, methane, and hydrogen sulfide.

Three separate experiments were performed by each of the landers. Two of the three experiments were designed to monitor and detect released gases from any microorganisms present. The gas exchange experiment proceeded by adding a nutrient solution to a soil sample. The experiment proceeded in two steps. First, enough of the nutrient was added to humidify the test container without wetting the soil. Carbon dioxide and oxygen increased by factors of 5 and 200, respectively. The soil sample was then saturated with nutrient. Carbon dioxide levels continued to rise but oxygen levels gradually decreased. These results were consistent with chemical reactions between a soil rich in oxides and the nutrient solution, not life processes.

The labeled release experiment employed a nutrient solution whose carbon compounds contained radioactive carbon 14 (^{14}C). Again, nutrient was added only in amounts sufficient to humidify the test chamber. Carbon dioxide gas containing ^{14}C was quickly released, however, the reactions were strictly chemical responses not associated with life forms.

The third experiment was designed to simulate the temperature range and atmospheric composition on Mars. Its primary function was to determine if organic matter had been synthesized after radioactive carbon dioxide and carbon monoxide were introduced into the chamber. A xenon lamp provided synthetic sunlight. At the end of the experiment, the soil was heated in order to break up any organic molecules into gaseous forms and then treated to produce carbon dioxide. The presence of radioactive carbon in the carbon dioxide would verify that organic compounds had been formed in the test chamber. This experiment was performed nine times. Seven positive results were obtained, however, these results are now suspect. It seems that some of the samples were stored for brief periods at $363^{o}K$ and others at $458^{o}K$. The sample results for the $458^{o}K$ samples showed a 90 percent decrease in end products.

Proponents for the discovery of life point to the positive results at lower temperatures. Arguments against the discovery of microorganisms concentrate on the higher temperature analysis. If life forms were present, they should have perished at temperatures of $458^{o}K$. The results indicated that some type of organic synthesis ($\approx$10% of the original amount) had continued. These results could only be explained by an unanticipated chemical reaction as the higher temperatures should have killed any microorganisms. Currently, the consensus of opinion seems to favor a nonbiological explanation. The question of life will not be resolved until Mars is revisited and reexamined, an unlikely prospect in the near future.

Summary

Mars is a desert planet with a rarified atmosphere that is mostly carbon dioxide. Water exists primarily in the vapor and crystalline states although subsurface liquid water is still a possibility. Surface winds are gentle, increasing with altitude. Condensation processes produce few clouds. The most significant meteorological events are the massive dust storms which envelop the planet in an opaque veil for weeks at a time. Occasional cyclonic storms have been observed near the polar caps' perimeters. Cyclic orbital variations provide arguments for dramatic long-period climatic changes. The changing Martian environments, in turn, provide fuel for the theories about the possibilities of life on Mars. To date, hard evidence for life on Mars is nonexistent. It may be decades before another scientific probe visits Mars. In the interim, we shall have to be content with the data payload returned by Viking 1 and 2.

References

Allen, C.C., Volcano-ice interactions on Mars, J. Geophys. Res., 84, 8048 ff., 1979.

Arvidson, R.E. et al., The surface of Mars, Scientific American, 238, 76 ff., 1978.

Briggs, G. et al., Martian dynamical phenomena during June-November 1976: Viking orbiter imaging results, J. Geophys. Res., 82, 4121 ff., 1977.

Briggs, G.A. et al., Viking orbiter imaging observations of dust in the Martian atmosphere, J. Geophys. Res., 84, 2795 ff., 1979.

Carr, M.H. and Schaber, G.C., Martian permafrost features, J. Geophys. Res., 82, 4039 ff., 1977.

Cutts, J., The polar regions of Mars, Astronomy, 5, 10 ff., 1977.

Farmer, C.B. et al., Mars: Water vapor observations from the Viking orbiters, J. Geophys. Res., 82, 4225 ff., 1977.

Farmer, C.B. et al., Mars: northern summer ice cap - water vapor observations from Viking 2, Science, 194, 1339 ff., 1976.

Farmer, C.B. and Doms, P.E., Global seasonal variation of water vapor on Mars and the implications for permafrost, J. Geophys. Res., 84, 2881 ff., 1979.

Gribbin, J., Martian climate: past, present and future, Astronomy, 5, 18 ff., 1977.

Hanson, W.B. et al., The Martian ionosphere as observed by the Viking retarding potential analyzers, J. Geophys. Res., 82, 4351 ff., 1977.

Hess, S.L. et al., Early meteorological results from the Viking 2 lander, Science, 194, 1352 ff., 1976.

Hess, S.L. et al., Mars climatology from Viking 1 after 20 sols, Science, 194, 78 ff., 1976.

Hess, S.L. et al., Meteorological results from the surface of Mars: Viking 1 and 2, J. Geophys. Res., 82, 4559 ff., 1977.

Hess, S.L. et al., The seasonal variation of atmospheric pressure on Mars as affected by the south polar cap, J. Geophys. Res., 84, 2923 ff., 1979.

Horowitz, N., The search for life on Mars, Scientific American, 237, 52 ff., 1977.

Hunt, G. and James, P., Martian extratropical cyclones, Nature, 278, 531 ff., 1979.

Jones, K. et al., One Mars year: Viking lander imaging observations, Science, 204, 799 ff., 1979.

Kieffer, H.H. et al., Martian north pole summer temperatures: dirty water ice, Science, 194, 1341 ff., 1976.

Leovy, C.B., The atmosphere of Mars, Scientific American, 237, 34 ff., 1977.

Levine, J.S., The evolution of water on Mars, in Proceedings of the colloquium of Water in Planetary Regoliths, Dartmouth College, Hanover, N.H., 34 ff., 1976.

Mantas, G.P. and Hanson, W.B., Photoelectron fluxes in the Martian ionosphere, J. Geophys. Res., 84, 369 ff., 1979.

McElroy, M.B. et al., Photochemistry and evolution of Mars' atmosphere: a Viking perspective, J. Geophys. Res., 82, 4379 ff., 1977.

Nier, A.O. and McElroy, M.B., Composition and structure of Mars' upper atmosphere: results from the neutral mass spectrometers on Viking 1 and 2, J. Geophys. Res., 82 4341 ff., 1977.

Owen, T. et al., The atmosphere of Mars: detection of krypton and xenon, Science, 194, 1293 ff., 1976.

Owen, T. et al., The composition of the atmosphere at the surface of Mars, J. Geophys. Res., 82, 4635 ff., 1977.

Pickersgill, A. and Hunt, G., The formation of Martian lee waves generated by a crater, J. Geophys. Res., 84, 8317 ff., 1979.

Pollack, J. et al., Winds on Mars during the Viking season: predictions based on a general circulation model with topography, Geophys. Res. Let., 3, 479 ff., 1976.

Pollack, J. et al., Properties of aerosols in the Martian atmosphere, as inferred from Viking lander imaging data, J. Geophys. Res., 82, 4479 ff., 1977.

Rasool, S.I. and Le Sergeant, L., Implications of the Viking results for volatile outgassing from earth and Mars, Nature, 266, 822 ff., 1977.

Sagan, C., Reducing greenhouses and the temperature history of earth and Mars, Nature, 269, 224 ff., 1977.

Seiff, A. and Kirk, D., Structure of Mars' atmosphere up to 100 kilometers from the entry measurements of Viking 2, Science, 194, 1300 ff., 1976.

Seiff, A. and Kirk, D., Structure of the atmosphere of Mars in summer at mid-latitudes, J. Geophys. Res., 82, 4364, 1977.

Snyder, C.W., The planet Mars as seen at the end of the Viking mission, J. Geophys. Res., 84, 8487 ff., 1979.

Snyder, C.W., The extended mission of Viking, J. Geophys. Res., 84, 7917 ff., 1979.

Thomas, P. and Veverka, J., Seasonal and secular variations of wind streaks on Mars: an analysis of Mariner 9 and Viking data, J. Geophys. Res., 84, 8131 ff., 1979.

Veverka, J. et al., A study of variable features on Mars during the Viking primary mission, J. Geophys. Res., 82, 4167 ff., 1977.

Yanagita, S. and Imamura, M., Excess ^{15}N in the Martian atmosphere and cosmic rays in the early solar system, Nature, 274, 234 ff., 1978.

CHAPTER V

JUPITER

Descriptive Statistics:

Radius	
Equator	71,400 km
Polar	66,800 km
Period of Rotation	9 hr, 55 min, 29.7 sec (System III)
Distance from Sun	5.2 A.U.
Mass	317 times that of the earth
Mean Density	1.33 g cm^{-3}

Introduction

The planet Jupiter appears to be an enigma to our solar system (Fig. 5.1). Its chemical composition more closely resembles the sun than any of the other planets. Cosmologically, this fact has generated a number of interesting ideas concerning the origin of our solor system and particularly, the formation of Jupiter. However, a discussion of these ideas must be deferred, out of necessity until we have explored the factual information currently available on Jupiter's atmosphere.

Atmospheric Composition

Numerous studies from ground based optical telescopes and Pioneer 10 satellite data have revealed that Jupiter's atmosphere consists of hydrogen (H_2), helium (He), ammonia (NH_3), methane (CH_4), phosphine (PH_3), water (H_2O), and possibly hydrogen sulfide (H_2S). The Voyager probes have also detected acetylene (C_2H_2), ethane (C_2H_6), and germanium tetrahydride (GeH_4). At Jupiter's temperatures and pressures, these would not be solids except at upper atmospheric levels, but rather exist in a gaseous or liquid state throughout the interior of the planet. It should be noted that the nature of Jupiter's deep interior is unknown. A solid core for Jupiter has neither been confirmed nor disproved as yet, although the evidence seems to be pointing toward a liquid core. Assuming a liquid core, we would expect convection to be the dominant process not only

Fig. 5.1. Voyager 1 photograph of Jupiter taken on January 29, 1979 from 35.6 million km. (Courtesy Jet Propulsion Laboratory/NASA.)

distributing heat from within the interior of the planet to the surface but also stirring the liquid and gaseous envelope of the planet in the process.

Hydrogen and helium account for 99 percent of the elements on Jupiter by volume. The ratio of hydrogen to helium is approximately 9:1 on Jupiter. This translates into an atmosphere of 88 percent hydrogen, 11 percent helium, and 1 percent all other elements. Based on the relative strengths of absorption spectra of hydrogen, methane, and ammonia, the relative abundance ratios of these gases has been determined. The ratio of carbon to hydrogen was 1:3,000, nitrogen to hydrogen 1:10,000, and helium to hydrogen approximately 1:9. The combination of the hydrogen/helium ratio and Jupiter's low density (1.33 g cm^{-3}) suggests that the planet is much more like the sun in composition.

Ethane (C_2H_6), acetylene (C_2H_2), phosphine (PH_3), and germanium tetrahydride (GeH_4) were recently discovered in Jupiter's upper atmosphere. Additional evidence has been found to suggest that hydrogen cyanide (HCN) may also be present. The presence of ethane and acetylene in the upper atmosphere was not unsuspected. These gases are probably produced as incoming solar ultraviolet radiation dissociates molecular hydrogen and organic molecules, particularly methane and ammonia. Although both ethane and acetylene are minor constituents of Jupiter's atmosphere, the abundance of ethane to acetylene (their abundance ratio) actually increased by a factor ≃2.0 during the four months intervening between Voyager 1 and Voyager 2. A hemispheric asymmetry in the abundance ratio of ethane/acetylene was also discovered with larger abundance ratios measured in the higher latitudes of the northern hemisphere. The volume of acetylene present in Jupiter's atmosphere also tends to decrease with increasing latitude.

Phosphine was discovered in Jupiter's atmosphere by scientists using high resolution interferometer spectrometer techniques in the 8-13 μm region. Its discovery triggered the development of several hypotheses concerning not only the photochemistry of phosphine but also its role in the red coloration of some of the cloud regions, particularly the Great Red Spot.

The presence of phosphine in Jupiter's observable atmosphere has led to a better understanding of both its photochemistry and the circulation of the planet. Models of Jupiter's atmosphere indicate that phosphine should exist only deep within the atmosphere at a temperature of $800^{o}K$ or greater, while between $800^{o}K$ to $300^{o}K$, phosphine would be oxidized to P_4O_6 by water and above the $300^{o}K$ level, the P_4O_6 would condense out to form aqueous H_3PO_3. The discovery of phosphine permitted Prinn and Lewis (1975) to develop a model to account for the red coloration based on the following photochemical and chemical reactions of phosphine with hydrogen:

$$PH_3 \rightarrow PH_2 + H$$

$$PH_2 + PH_2 \rightarrow PH + PH_3$$

$$PH + PH \rightarrow P_2 + H_2$$

$$H + H + M \rightarrow H_2 + M$$ (where M is any background molecule)

$$P_2 + P_2 + M \rightarrow P_4 + M$$

$$P_4 \rightarrow P_{4(s)}.$$

For temperatures between $290^{o}K$ to as high as $600^{o}K$ and pressures ranging from 0.1-1,000 mb, the above sequence of reaction appears valid with the exclusive end products being hydrogen (H_2) and solid red phosphorus ($P_{4(s)}$). The solid particles then move downward by turbulent diffusion and gravitational sedimentation to the 20-bar level where they quickly evaporate to gaseous phosphorus (P_4). Below the $800^{o}K$ level the cycle is completed by the following thermo-chemical reaction: $P_4 + 6H_2 \rightarrow 4PH_3$ which is relatively fast compared with the vertical transport rates. Computer modelling studies prompted Prinn and Lewis to conclude that the coloration of the Great Red Spot is a natural consequence of strong vertical mixing in this particular region. It would therefore seem logical to conclude from their (Prinn and Lewis) arguments that the red phosphorus ($P_{4(s)}$) produced from the dissociation of phosphine is the primary red chromophore in Jupiter's atmosphere. It is therefore also reasonable to assume that the redness of a particular region is a function of dynamic, convective mixing. The Great Red Spot (GRS) would be the ultimate example of such a region in Jupiter's southern hemisphere.

Contrary ideas explaining the red colors on Jupiter exist. Photographic evidence has detected what appear to be massive, convective, cumuliform clouds with bright white tops. Since vertical motions are associated with both the GRS and these cumuliform-like plumes, it has been suggested that the colors of the clouds are a function of chemical reactions whose rates are controlled by the rapidity and distance of the vertical ascent.

An alternative theory proposes that polymerized carbon suboxide $(C_3O_2)_x$ is the red chromophore. Light polymerizes the liquid and vapor form of carbon suboxide into a red solid. The carbon and oxygen to form carbon suboxide exist on Jupiter as by products of the photodissociation of atmospheric gases. Perhaps carbon suboxide, rather than phosphine, is transported vertically within the GRS. It would be polymerized by the sunlight producing a solid, red aerosol substance which would slowly precipitate out, coloring the upper cloud tops of the GRS in the process. The actual detection of phosphine suggests that it is primarily responsible for the reddish cloud colors, however, until definitive proof is abtained, we should not totally discount the alternative proposals.

The presence of germanium tetrahydride (GeH_4) and the possibility that hydrogen cyanide (HCN) has initiated speculation that at least some elementary life forms might exist on Jupiter. While both gases are toxic, they are the chemical building blocks for other organic compounds. R. Dimmick, Chairman of the Aerosol Science Group at the Naval Biosciences Laboratory in Oakland, California, reported that microorganisms may be capable of living in a gaseous suspension, i.e. Jupiter's atmosphere, without either solid surfaces or a liquid. Working with the earth bacterium, Serratia marcescens, Dimmick's group discovered that these microbes could survive in an aerial suspension by metabolizing sugar and producing new DNA. Some evidence was also found that reproduction was occurring in the aerial suspension. Serratia marcescens is an oxygen breathing bacterium. Oxygen is not present in Jupiter's atmosphere but there does not seem to be any reason why other organisms, attuned to Jupiter's atmospheric chemistry, might not exist. The recent discovery of bacteria in seeping springs in northern California, which survived environments ten times more alkaline than previously thought possible destroys one of the, heretofore, primary arguments against life on Jupiter. It had been argued that the alkalinity of Jupiter's atmosphere, due to ammonium hydroxide (NH_4OH), would render it inhospitable to life. Now, at least we know that life forms may exist in the gaseous, atmospheric envelope of Jupiter.

Carbon monoxide (CO) and carbon dioxide (CO_2) have not been detected on Jupiter. There is some evidence to suggest that carbon monoxide molecules may be present in the ratio of one carbon monoxide molecule to one billion hydrogen molecules. The presence of gaseous carbon dioxide would be impossible to explain. The known reactions between carbon dioxide and a hydrogen rich atmosphere would preclude the existence of CO_2. Thus, Jupiter would be the first planet we have examined whose atmosphere is devoid of CO and CO_2.

The Structure of Jupter's Atmosphere

The Jovian atmosphere is divided into four major layers: troposphere, stratosphere, mesosphere, and thermosphere. The troposphere extends from somewhere within the planet to a pressure of 100 mb or 0.1 earth atmospheres. A well defined tropopause separates the troposphere from the stratosphere. The stratosphere extends from the pressure levels of 100 mb to 0.1 mb. The pressure heights of the mesosphere and thermosphere have not been accurately delimited as yet.

The known temperature/pressure relationships are presented in Fig. 5.2. Temperatures decrease with height to a well-defined

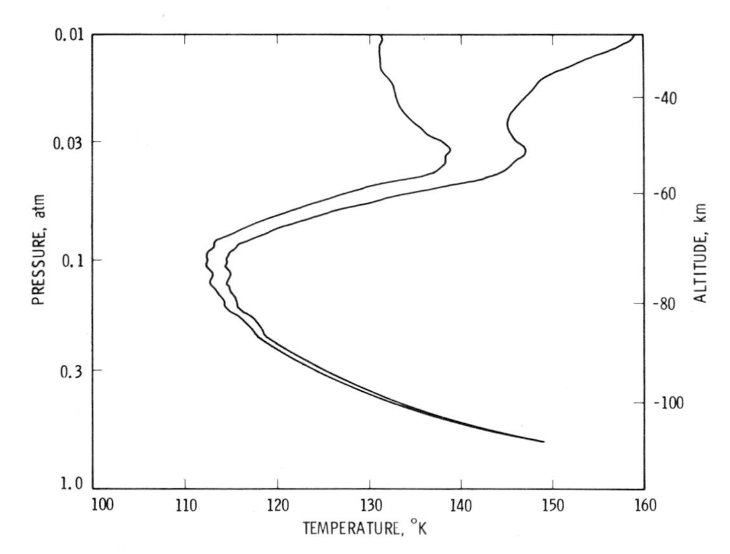

Fig. 5.2. The vertical temperature/pressure structure from Voyager 1 data. (Courtesy V.R. Eshleman et al., Science, v 204, p. 977, 1 June 1979, Copyright 1979 by the American Association for the Advancement of Science.)

tropopause. Temperatures increase in the Jovian stratosphere, however, not for the same reasons as temperatures increase in the earth's stratosphere. The temperature increase is probably due to the absorption of solar energy by a layer of methane along with ethane, acetylene, and phosphine. This warming is analogous to the warming of the earth's stratosphere through the absorption of ultraviolet energy by molecular (O_2) and triatomic oxygen (O_3). The data depicting temperature trends in the mesosphere and thermosphere are very sketchy. Occultation data have revealed a decrease of temperature with height in thc mesosphere and a sharp increase in temperature in the thermosphere. The temperature decrease in the mesosphere and increase in the thermosphere are analogous to those of earth's atmosphere and may be explained by the same reasons given for those layers in chapter one. The ionosphere extends more than 3,000 km above the 1.0 mb level. This depth is thought to be the result of the diffusion of ionized gas at very high temperatures. The upper ionospheric electron-ion temperature was originally reported by the Pioneer probe at about 870°K. The Voyager probes recorded much larger ionospheric temperatures ranging from 1,100^{o}-1,300^{o}K. There are indications that the Jovian ionosphere has at least several well-defined layers. Some researchers suggest the possibility of as many as seven layers based on indications of differing electron densities at different altitudinal levels. The second layer from the top seems especially pronounced and may be the product of solar ultraviolet radiation. It is necessary to postpone our discussion of the ionosphere until we have discussed Jupiter's magnetic characteristics so that we may understnad the nature of the ionosphere and its relationship to the rest of the atmospheric structure.

A detailed vertical temperature/pressure diagram of the troposphere and lower stratosphere at three select locations, 10°N, 15°S, and the GRS (23°S) is presented in Fig. 5.3. The middle and lower troposphere (300-600 mb) is relatively uniform with temperatures decreasing with altitude at adiabatic or near adiabatic rates of 2°K km^{o1}. Significant variations begin to appear in the upper troposphere and are readily apparent at the tropopause. The tropopause temperatures for 15°S, 10°N, and the GRS were ≈112°K, 105°K, and 100°K, respectively. The temperatures at 15°S were warmer and the height of the tropopause was lower. Temperatures for all three sites depict a steady increase in temperature with height in

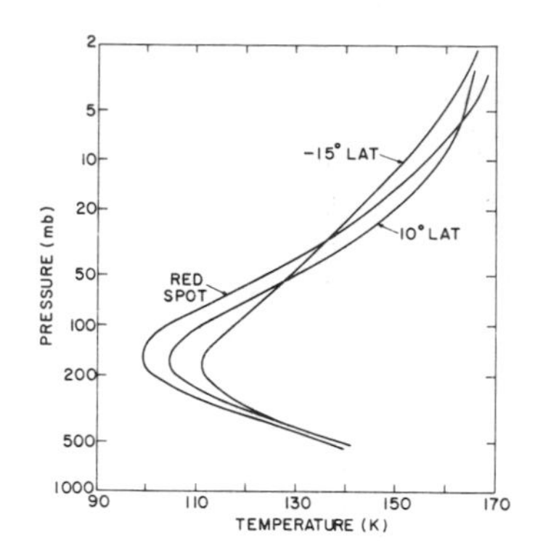

Fig. 5.3. Three vertical temperature/pressure profiles for 10°N, 15°S and the Great Red Spot (20°S). (Courtesy R. Hanel et al., Science, v 204, p. 973, 1 June 1979. Copyright 1979 by the American Association for the Advancement of Science.)

the stratopause. Radio occultation data from another experiment reported the existence of a strong inversion in the Jovian stratosphere at 35 mb. This inversion might delineate a boundary, possibly the upper haze boundary, that is thought to exist somewhere near this level, if it exists at all.

A vertical-latitudinal, cross-section depicting the thermal structure of Jupiter's atmosphere shows relatively level isotherms at the tropopause with warmest tropopause temperatures at 15°S (Fig. 5.4). The clouds associated with the belt and zone pattern apparently exert smaller thermal effects than had been anticipated. The strong latitudinal orientation of the stratospheric isotherms was unexpected. Scientists were expecting a more meridional type of circulation pattern in the stratosphere which would, had it existed, have shown strong equator to pole temperature variations in the isothermal patterns.

A Voyager 1 isothermal map for most of Jupiter corresponding to an altitude of 150 mb and presented as a cyclindrical projection depicts the thermal structure of the tropopause region (Fig. 5.5). It is evident that the generally belt-like cloud patterns are present because of the relative uniformity of the isothermal profile. The cold regions located at 20°-35°N and 20°-35°S are readily visible. These are synonymous with the bright zones seen in Voyager 2 color

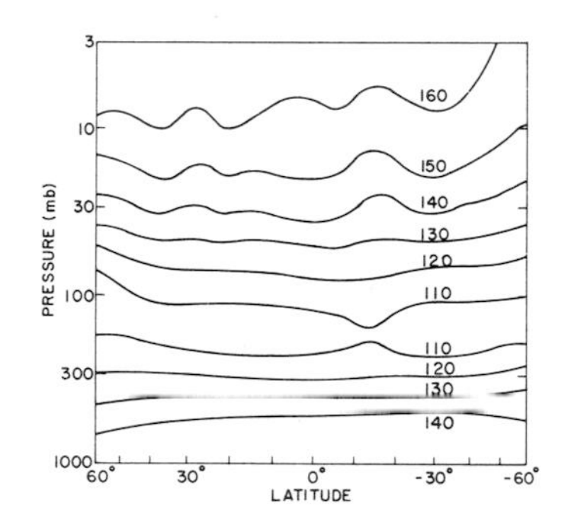

Fig. 5.4. Latitudinal isothermal cross-section of Jupiter's atmosphere. (Courtesy R. Hanel et al., Science, v 204, p. 973, 1 June 1979. Copyright 1979 by the American Association for the Advancement of Science.)

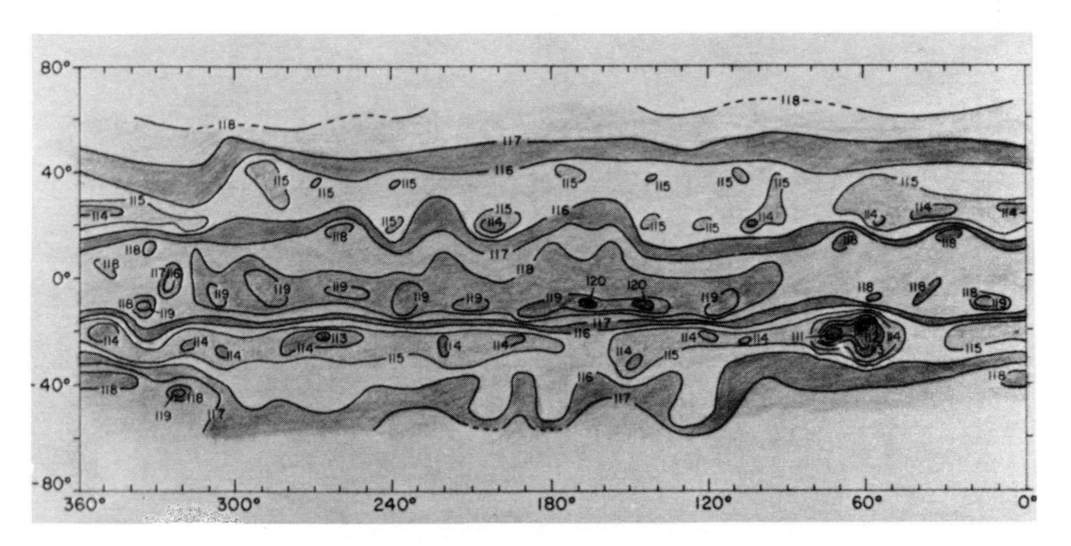

Fig. 5.5. Voyager 1 brightness temperature map corresponding to an altitude of 150 mb. (Courtesy R. Hanel et al., Science, v 204, p. 955, 1 June 1979. Copyright 1979 by the American Association for the Advancement of Science.)

photography. The warmest regions on this map straddle the equator from 15°N to 15°S. This area corresponds with the darker equatorial region in the color images. Cold features such as the GRS (23°S, 105°W) are readily apparent as are warm features (18°S, 240°W) which correspond to the darker brown elliptical areas on the color plates. It is therefore possible to not only measure and map atmospheric temperatures for selected pressure levels but also to correlate the patterns that appear on these isothermal maps with visible cloud patterns. The information obtained by comparing this Voyager map to a similar map produced from Voyager 2 data, revealed that changes had occurred in the cloud patterns during the intervening four month period. Prior to these observations it was thought that Jupiter's cloud patterns changed very slowly, perhaps over a several year or more time span. Recognition of changes that had occurred in four months suggested dynamic processes at work in the atmosphere of this gaseous giant.

The Magnetic Field and Magnetosphere

Jupiter, like the earth, has an intrinsic planetary magnetic field (Fig. 5.6). The origin of Jupiter's magnetic field probably lies deep in the planet's interior where electric arcs are created by Coriolis deflection of convection currents within the region of liquid, metallic hydrogen surrounding a rocky core. It is thought that this same process may be responsible for generating the earth's magnetic field within the liquid iron and nickel outer core region. Since the origin of planetary magnetic fields is not well understood, any similarities and/or comparisons of the origin of planetary magnetic fields should be considered very speculative.

The Jovian magnetic field is primarily dipolar in nature with north and south magnetic poles. These poles are displaced from the geographic poles and rotational axis by ≈10° (Fig. 5.6). This 10° offset from the rotational axis combined with Jupiter's rapid rotation rate of 9 hours, 55 minutes, 29 seconds causes the magnetic field and the radiation belts associated with it to wobble up and down through a 20° arc every day. The field strength, which is approximately 10 times greater then the earth's field strength, also varies by 400-500 percent, due to this wobbling effect.

The shape of the magnetosphere is highly asymmetrical (Fig. 5.6). The portion of the magnetosphere facing the solar wind is compressed

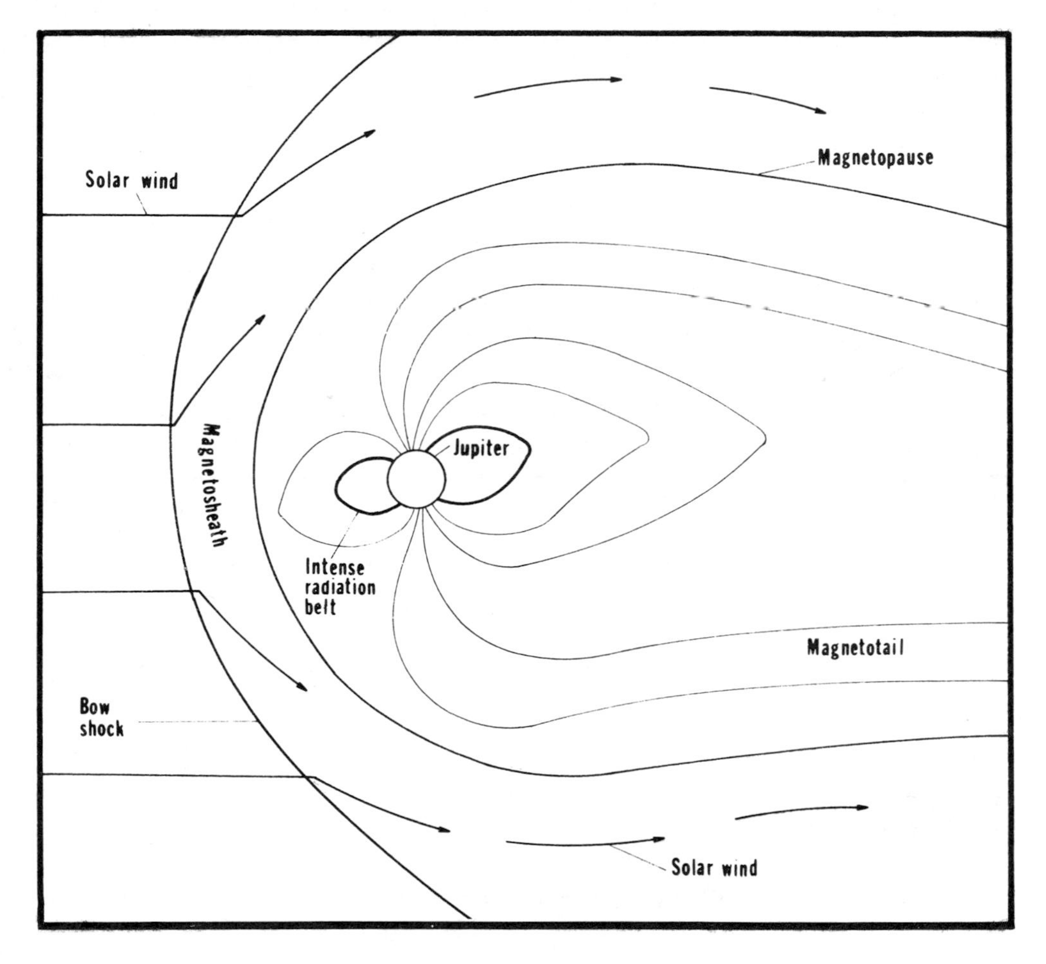

Fig. 5.6. Jupiter's asymmetrical magnetosphere, magnetopause, and bow shock region. Two intense radiation belts surround Jupiter.

while the antisolar side exhibits a massive magnetotail. The boundary between the magnetosphere and the solar wind is complex, consisting of the magnetopause, magnetosheath, and bow shock regions. The bow shock region is identified primarily by a change in the particle flow direction of the solar wind caused by the clash of the solar wind with Jupiter's magnetic forces. The solar wind particles are not only deflected around Jupiter but decelerate from a velocity of 1,500,000 km hr^{-1} to less than 500,00 km hr^{-1}. The magnetosheath is a buffer region between the bow shock and the magnetopause, the outer.limit of the magnetic field. The Voyager probes recorded an unusual sequence of events when they first crossed the bow shock at a distance of ≈99 Jupiter radii (R_J). They later crossed and recrossed the bow shock until they were within 62 R_J of the planet. The magnetosphere of Jupiter had actually collapsed and expanded like an accordian due to compression from the ram pressure of the solar wind. Variations in ram pressure result from disturbances on the sun affecting the acceleration of solar wind particles. The region between 60 R_J and the planet does not appear to experience bow shock passages. The region of the magnetosphere within 20 R_J would be typified as a classic magnetosphere where extremely high energy particles are trapped and oriented in intense radiation belts similar to, but 5,000-10,000 times more intense than the earth's Van Allen Belts. The antisolar side of Jupiter appears somewhat more tranquil as the solar wind helps to shape a magnetotail extending between 300-400 R_J away from the planet.

The entire magnetosphere is a plasma region, i.e. a region of significant ion density. The combined effects of Jupiter's powerful magnetic field and rapid rotation are analogous to a giant particle accelerator where electrons, protons, and heavy ions of the plasma population are accelerated and segregated. The vast majority of the plasma population aligns itself along the magnetic equator in a magnetodisk. The central region of the magnetodisk carries a directional electric current termed the current sheet.

The compression of the magnetospheric plasma is so intense that electrons are accelerated to velocities approaching the speed of light. Many of these particles are ejected from Jupiter altogether, some arriving in the vicinity of the earth. Thus, it would seem that both Jupiter and the sun are important sources of highly energetic particles in the solar system. Understanding how Jupiter's magnetosphere accelerates particles actually helped to resolve a scientific mystery. It had been reported that the Pioneer spacecraft had detected high energy electrons moving toward the earth in advance of the Jovian bow shock. Once the process of particle ejection was understood, a reexamination of previously gathered earth satellite data was undertaken, concentrating on the information showing an anomalous increase of cosmic background level electrons every 13 months. It was recognized that every 13 months, both earth and Jupiter are connected by the lines of the interplanetary magnetic field. In space, high speed electrons travel along the magnetic field lines. The detection and source of these high speed electrons had puzzled scientists. Now, we know that these high speed electrons which arrive in the earth's vicinity came from Jupiter's magnetospheric plasma.

Jupiter's magnetospheric plasma is extremely complex. Plasma temperatures are high from the outer edge in to 30 R_J ranging from 3×10^{8} °K to 5×10^{8} °K with particle densities of 5×10^{-3} cm^3. Jupiter's four largest moons, Io, Ganymede, Europa, and Callisto all orbit within the magnetospheric plasma. Io's orbit around Jupiter (Fig. 5.7) is described as a plasma torus. The torus is a

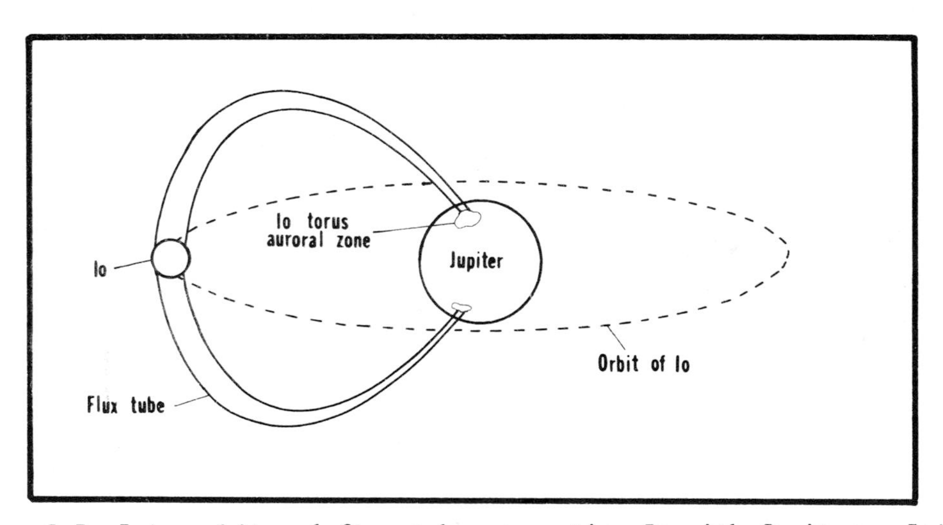

Fig. 5.7. Io's orbit and flux tube connecting Io with Jupiter. Io's torus is a doughnut-shaped ring symmetrical with Io's orbit.

doughnut-shaped path within which Io orbits. Hot plasma temperatures exist within this torus decreasing in all directions away from the torus. The plasma, particularly within 20 R_J, is composed of ionized forms of sulfur, helium, oxygen, carbon, hydrogen, and sodium. There is a distinct possibility that some heavy molecular ions such as sulfur, sulfur dioxide, zinc sulfide, and sulfur salts or polymers may also be present. The carbon, hydrogen, and helium probably originate from the solar wind or have been accelerated out of the Jovian ionosphere. Io, the closest of the four large moons, is the source of the remaining ions, which comprise the bulk of the plasma. The mechanisms which remove materials from Io's surface and distribute them in the torodial path are volcanic activity and sputtering, a process where the collision of high energy particles with Io's surface propels materials spaceward at the necessary escape velocity. We shall discuss these processes more thoroughly in the chapter concerning Io.

The plasma disk and Io torus are not the only intriguing magnetospheric characteristics. The rapidly rotating magnetic field lines penetrate and sweep past Io generating the powerful electric current flowing between Jupiter and its moon. The result is the creation of the Io flux tube (Fig. 5.7) where observations indicate voltages in excess of 400,000 volts and a 5,000,000 ampere current flow between Io and Jupiter's upper atmosphere. The electric power generated in this flux tube is in excess of two trillion watts or twenty times the total combined generating capacity of all the earth's nations. These currents "touch" the Jovian ionosphere at the foot of the flux tube, producing significant auroral displays and decametric radio emissions. These topics will be discussed later in this chapter.

A new magnetic phenomon has been observed in association with Jupiter's magnetotail. The rotation of the plasma is so strong that on the antisolar side, the Voyager probe detected plasma breaking away from the corotational flow and moving away from Jupiter in the same direction as the magnetotail. This breaking away of plasma appears as a continuous flow moving away from Jupiter at 300-1,000 km sec^{-1} and has been termed the magnetospheric wind.

Energy Budget

The energy budget for Jupiter and earth represent two different heat absorption and radiation processes (Fig. 5.8A & B). The solar energy absorption curves indicate a relatively uniform pattern on both planets. The maximum amount of solar energy is absorbed in the

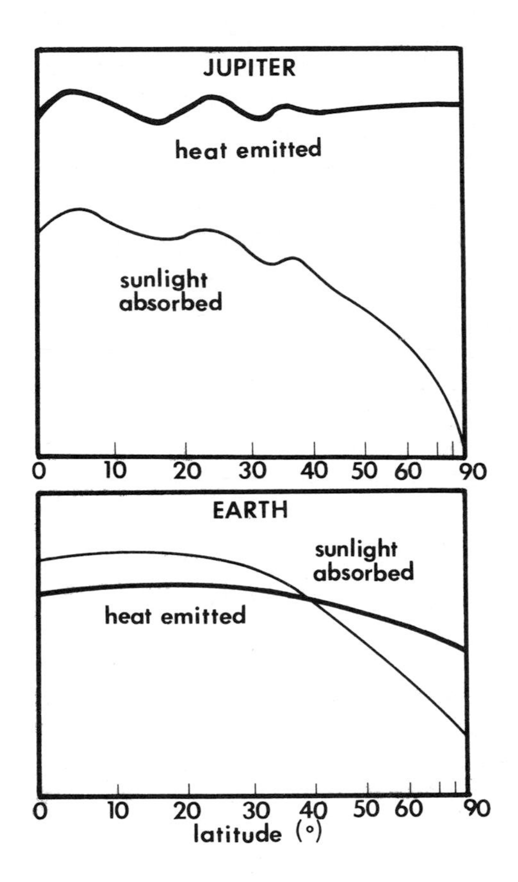

Fig. 5.8. Energy absorption and emission curves for the earth and Jupiter. (Courtesy Space World, v m-9-153, p. 4, Copyright Palmer Publishing Co.)

equatorial region between 25°N and 25°S. The angle at which the sun's rays strike the planets' polar regions results in the absorption of less energy per unit area. The result, as depicted in the solar energy absorption curve, is that less energy is received and absorbed near the poles. The heat emission curves for the earth reveal that less energy is radiated from the equatorial regions than was absorbed. The emission curve for the polar latitudes shows that more energy is radiated from these latitudes than is received from the sun. The source region for this surplus energy which radiates out to space from the polar latitudes is the equatorial and subtropical regions. The heat energy is transported poleward by the warm ocean currents and by latent heat transfer associated with the earth's atmospheric circulation.

Jupiter's absorption curve reveals humps due to the differing absorption characteristics of the belt and zone regions. Yet, the equatorial-polar temperature range differs by as little as $\pm 3^{o}$K. What is more intriguing is that Jupiter's emission curve is relatively uniform for all latitudes and shows that Jupiter actually radiates more energy to space than it receives from the sun.

Pioneer data revealed that most of the solar energy absorbed by Jupiter's equatorial regions remains there and is reradiated out to space from these same latitudes. The high rotational velocity on Jupiter creates a zonal atmospheric circulation which is responsible for retaining the solar heat in these latitudes. On earth, rotating cyclonic storms (hurricanes/typhoons) are born between 5°-8° north and south latitudes. The absence of any significant Coriolis effect near the equator prevents the formation of rotating cyclonic storms and any associated latent heat transfer into subtropical or middle latitude regions. The mid-latitudes are the primary region for cyclonic storms on earth. Cyclonic vortices are much more common poleward of 45°-50° on Jupiter, as on earth. Convection on earth is a major energy exchange process. Convection is responsible for many of the 1,800-2,000 thunderstorms always in progress in the earth's atmosphere. Convection was originally not thought of as a significant energy exchange process on Jupiter but detailed Voyager images revealed clouds whose morphology was similar to earth cumulonimbi. Also, the 2°K lapse rate in the troposphere of Jupiter supports the possibility of convective activity. However, convection does not explain nor does it appear to be anywhere near important enough to be responsible for the observed polar infrared emissions.

The infrared emissions from Jupiter's polar regions are apparently a product of the planet's own internal heat source. The existence of two different sources of infrared emissions does explain the relative uniformity of thermal conditions from the equator to poles. The amount of energy Jupiter radiates above what it receives from the sun is the Jovian constant. The Jovian constant is 1.9 although earlier studies had estimated its value as high as 2.5. It is interesting to note that Saturn radiates 2.6 times more energy to space than it receives from the sun. The burning issue at present is to explain how a planet located 5.2 A.U. from the sun is able to radiate more energy than it intercepts from the sun. Jupiter's atmospheric composition of hydrogen-helium in a 9:1 ratio has led theorists to suggest that Jupiter might be an evolutionary form of a low mass star. The astronomical sequence of stellar matter collapse, ignition (the birth of a star), and entrance onto the main sequence of the Hertzsprung-Russell (H-R) stellar evolution diagram is still not well understood. As our solar system formed, the bulk of the material may have contracted and collapsed to ignite the sun, however, a similar, but secondary object like Jupiter was also formed.

Without a critical mass of hydrogen and helium, Jupiter never contracted enough to produce the extreme temperatures (15,000,000^{o}K) necessary for hydrogen (deuterium) conversion to helium. At some point in its collapse, the atoms simply refused to be crowded together any farther and contraction terminated leaving a giant gas planet with a very hot interior. The present day surplus of infrared radiant energy may come from an internal reservoir of thermal energy generated as Jupiter contracted as the solar system was being born. It is interesting to raise the question, "What changes would have resulted if the earth's evolutionary history had occurred in a binary star solar system?"

Atmospheric Pressure

The atmospheric pressure structure of Jupiter's atmosphere was presented in Fig. 5.2 as an integral part of the discussion on the vertical temperature structure. The measurement of atmospheric pressure is hindered by the absence of a planetary surface or reference level from which measurements may be made. Instead, pressure levels are referenced to temperature levels with the bottom of Jupiter's atmosphere arbitrarily defined as the point where atmospheric pressure equals 100 (earth) atmospheres. The bulk of atmospheric interactions, i.e. energy absorption, cloud bands, wind, occur in the region from 1.0-0.01 atmospheres (1,000-10 mb). The temperature at 1.0 atmosphere is ≃165°K. The tropopause temperature at 0.1 atmosphere is 113°K and varies greatly at 0.01 atmospheres because this upper stratospheric region is influenced by the hot Jovian plasma (800°-1,300°K).

Meteorological Phenomena

Cloud Regions

Jupiter's cloud and wind patterns depict a predominantly zonal distribution of phenomena consisting of zones, belts, and the Great Red Spot (and other related ovals). The global distribution of cloud regions and nomenclature are presented in Fig. 5.9. The EZ or equatorial zone, spans the equator parallelled by the NEB, North Equatorial Belt and SEB, South Equatorial Belt. Poleward of the NEB and SEB are the NTrZ, North Tropical Zone and the STrZ, South Tropical Zone. The NTeZ, North Temperate Zone and the STeZ, South Temperate Zone are poleward of the NTrZ and STrZ. Lastly, poleward of 45°-50°, the zonal pattern is replaced by a highly varied region of less orderly structure.

The cloud bands whose names end with the word "zone," i.e. the EZ, are generally characterized as areas of brighter (often white) colors (Fig. 5.1). The zones appear to be regions where upward motions are sustained. In contrast, the cloud bands whose names end with the word "belt," i.e. the NEB, are generally darker and browner and thought to be areas where the upper clouds are absent so that what we observe are lower, darker clouds, deeper in the atmosphere. The absence of the overlying clouds may be a result of subsidence in the belt regions.

There is some symmetry to the width of the cloud bands. They are wider in the equatorial latitudes because the Coriolis effect is negligible there. The width of the cloud bands poleward from the equator is inversely proportional to the magnitude of the Coriolis effect.

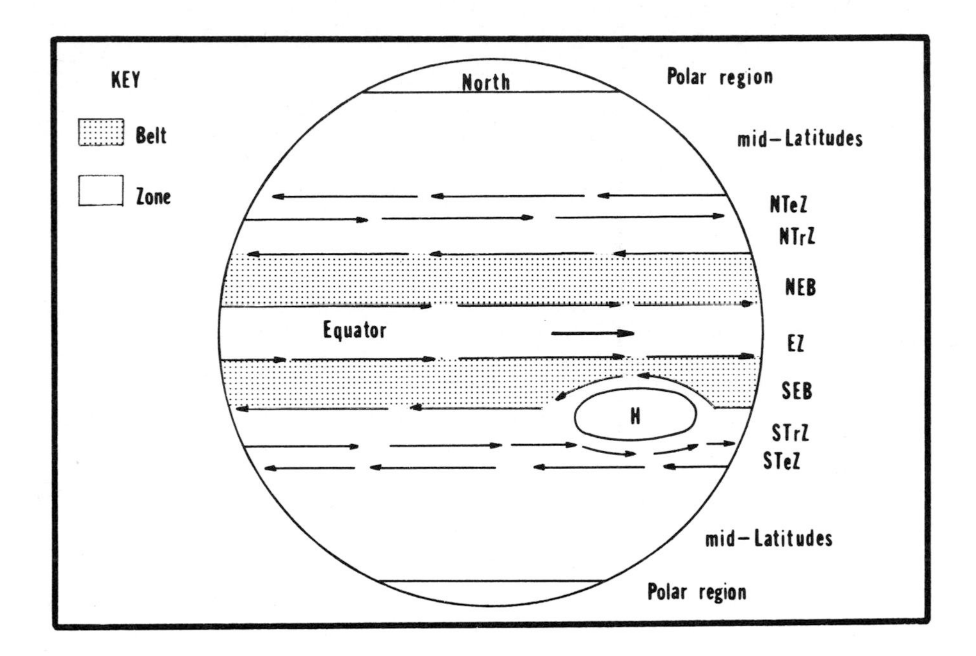

Fig. 5.9. The belts and zones of Jupiter. These features do not extend into the disturbed mid-latitude and polar regions. Compare this figure with Fig. 5.1. The arrows represent high velocity winds located at the zone/belt boundaries.

The Great Red Spot (GRS) is an important feature in any discussion of Jupiter's clouds. It is located at 23°S and is the single most striking feature of Jupiter's southern hemisphere. The GRS is similar to a zone because it represents a region of vertical atmospheric motion. The entire GRS rotates anticyclonically, i.e. counterclockwise in the southern hemisphere, and exhibits lower cloud top temperatures, indicative of very high clouds.

The general characteristics of the belts, zones, and GRS are summarized in Table 5.1. The belts, zones, and GRS should be considered quasi-permanent characteristics of Jupiter's atmosphere. The relative permanence or longevity of cloud features on Jupiter is probably due to the long radiative time constant, i.e. the time it takes for a mass of gas to warm or cool through infrared radiation. The radiative time constant on earth is but a few weeks whereas,

TABLE 5.1

The characterists of the belts, zones, and GRS.

Feature	Characteristics
Belts	- high thermal temperatures, low cloud height, cyclonic rotation, region of subsidence, and dark colored clouds
Zones	- low thermal temperatures, high cloud height, anti-cyclonic rotation, vertical motions upward, and light colored clouds
GRS	- low thermal temperatures, high cloud height, anti-cyclonic rotation, vertical motions upward, and thick reddish-orange clouds.

on Jupiter, it is more than a year and possibly much longer. The difference between the two planets is a result of Jupiter's lower atmospheric temperatures. Jupiter only receives 4 percent of the total solar radiation the earth receives. This lower energy receipt is further affected by the composition of the Jovian atmosphere. These gases emit less infrared radiation than those of the earth's atmosphere. Furthermore, it is also likely that the mass of gas in Jupiter's atmosphere stores its absorbed heat at low altitudes near the cloud base. If so, then the radiative time constant may be measured in years because of the huge pool of gas that must be cooled. Therefore, it is safe to assume that the explanation for the longevity of atmospheric phenomena is largely the result of the slow rate at which thermal energy is radiated out to space.

The winds on Jupiter also exhibit a zonal (E-W) pattern (Fig. 5.9). The highest velocities recorded were 130-150 m sec^{-1}. Typically, the highest velocity winds were located at the interfaces between the belts and zones. Table 5.2 presents the latitudes of eastward and westward flowing currents on Jupiter. There is little doubt that many of these currents exceed the fastest jet streams on earth as they literally scream around Jupiter.

The general cloud patterns of belts, zones, and GRS reveal only the largest scale features in Jupiter's atmosphere. A more detailed hemispheric analysis based on the Voyager results reveals far more detail and structure than had ever been anticipated.

Northern Hemisphere Clouds

There is a distinct asymmetry to the northern hemispheric cloud patterns when compared to those of the southern hemisphere. The

TABLE 5.2

Latitudes of eastward and westward currents with selected eastward velocities (after Smith et al., Science, v204, p952. Copyright 1979 by the American Association for the Advancement of Science)

Westward Currents	Eastward Currents	Velocity (m sec^{-1})
50N		
	47N	
44N		
	41N	
38N		
	35N	
30N		
	23N	150
18N		
	8N	-20
	0	120
	8S	90
16S		
	23S	130
28S		
	32S	-50
35S		

features in the northern hemisphere are generally smaller though similar in morphology to their southern hemisphere counterparts. The northern hemisphere also lacks any large structures such as the GRS. The diffuse clouds of the EZ are quite unlike the varied cloud patterns in the NEB (9°-18°N). This is a region of strong wind shear between an eastward flowing current at 8°N and a westward current at 18°N. The clouds are elongated and exhibit a filament-like structure. Small regions of bright cumuliform-like clouds, 500-2,000 km in diameter have been observed between 40°-100°W (within the NEB). They form and disappear in less than a few rotations of the planet. Along the northern boundary of the NEB between 140°-280°W, elongated, elliptical, dark brown structures exist. These brown spots are cyclonic features with very distinct boundaries. Thermal emissions at 5 μm show they are warmer than the surrounding regions (Fig. 5.10) suggesting that the high clouds may be absent, providing a view of the darker clouds deeper in the atmosphere. A close-up of one of these brown spots is presented in Fig. 5.11.

The region between the opposing currents at 18°N and 23°N is a bright area where the clouds appear linear but slightly off-axis to the wind shear. The standard nomenclature has been to refer to these clouds as having a chevron shape. The clouds at 31°N exhibit

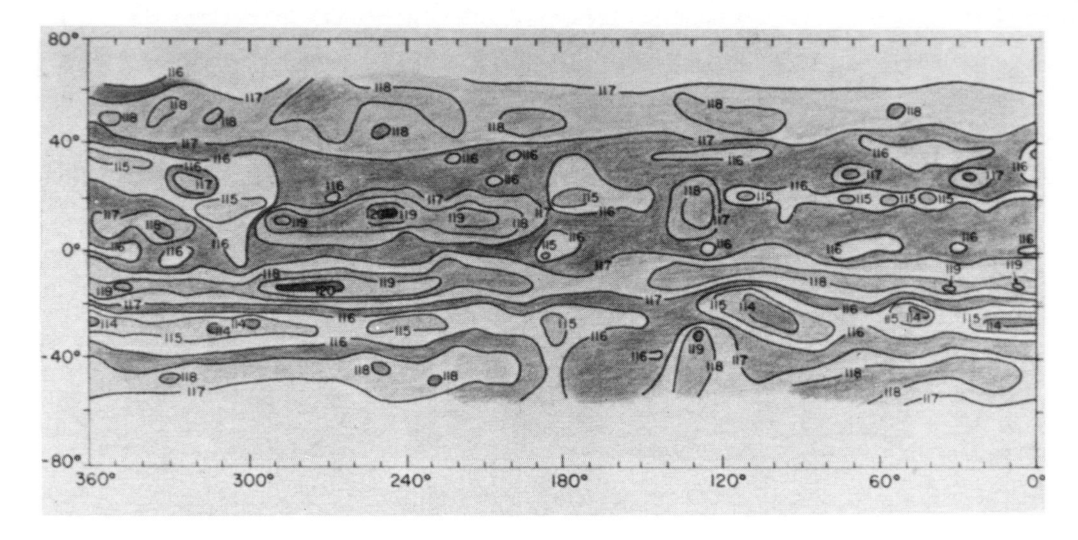

Fig. 5.10. Voyager 2 brightness temperature map of Jupiter corresponding to a pressure level of 150 mb. (Courtesy R. Hanel et al., Science, v 206, p. 954, 23 Nov. 1979. Copyright 1979 by the American Association for the Advancement of Science.)

a pattern similar to a piece of folded ribbon while at 35°N, a transition area with circular patterns ≃3,000 km in diameter exists. These circular features exhibit clockwise rotation, bright central regions, and dark outer rings. There is a rough symmetry to these features as they are spaced 10°-15° of longitude apart. They are found in a distinct longitudinal region 350°-260°W. They are totally missing from the area between 260°-350°W.

The region from 36°-80°N is a mottled area characterized by a number of white, circular features with interior, spiral cloud filaments rotating in a clockwise pattern. Large areas of randomly distributed cumuliform-like clouds and regions of folded ribbon clouds populate this broad latitudinal span. Recirculating cyclonic currents have also been observed flowing east then turning north, west, and finally southward. The polar zones (80°-90°) have not been scrutinized by the various probes because the probes' trajectories have been at or near the planet's equatorial plane.

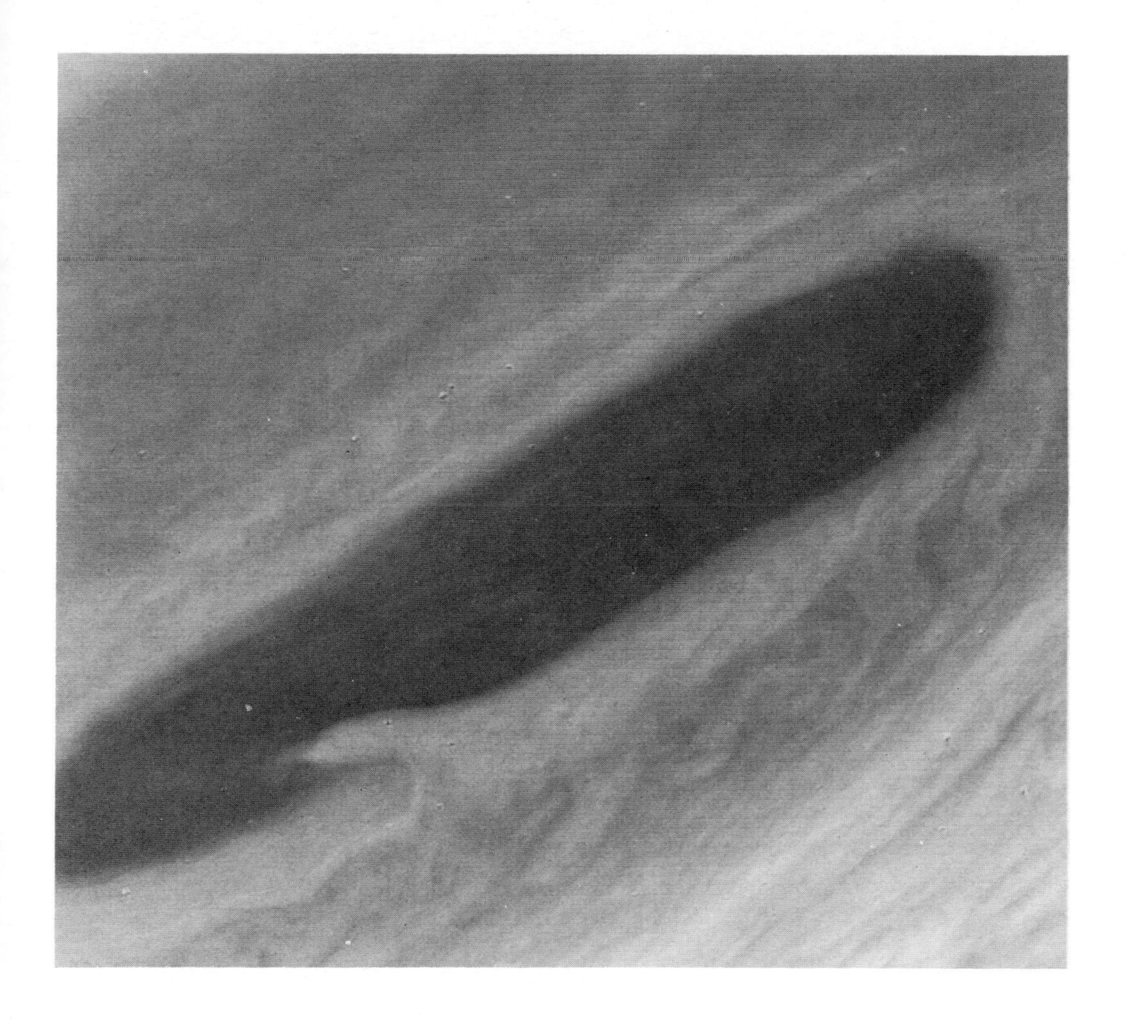

Fig. 5.11. Voyager 2 image of one of the long, dark clouds observed in the NEB taken on July 6, 1979 from 3.2 million km. Thin white clouds are also seen within the dark cloud. (Courtesy Jet Propulsion Laboratory/NASA.)

The Equatorial Zone (EZ) Clouds

The EZ is characterized as the darker zone between two eastward currents at 9°N and 9°S. Voyager I discovered a train of plumes suggestive of earth cumulonimbi strung out along the northern boundary of the EZ (Fig. 5.12). These plumes moved eastward at velocities of 100-150 m sec^{-1} though their movement was not uniform. Several of the plumes were characterized as bright and strongly suggest that this is an area of strong convective activity.

High resolution images revealed two types of wave phenomena in the EZ, plume tail waves and cross equatorial waves. The plume tail waves had axes perpendicular to the plume tails and were located 10,000 km from the actual plume. The plume tail waves appeared as light and dark regions spaced 300-650 km apart and were best developed as the plume center grew. These plume waves are caused by rapidly varying rates of convective ascent in the plume nucleus. They have also been observed on earth downwind of severe thunderstorms.

The cross equatorial waves were observed south of a well developed plume and extended longitudinally for 20,000 km. There is no known source mechanism for these waves as yet.

The Southern Hemisphere Clouds

Jupiter's southern hemisphere is very similar to the northern hemisphere. Eastward and westward jets alternate. Their width and magnitude decrease poleward. At 14°S, 31°S, and 40°S elongate cyclones alternate with elliptical anticyclones. The cyclones are not well defined in comparison to the elliptical anticyclones. There are three groups of elliptical spots, the GRS, white ovals, and spots at 41°S. These features all lie just south of a westerly jet, no doubt receiving some of their rotational energy from the shear at the westward jet boundary. The elongate cyclones lie just to the north of these jets thus receiving cyclonic rotational energy.

Certain characteristics are common to the GRS, white ovals, and 41°S spots. They are anticyclonic whirls and lack an organized structure. Chains of cumuliform-like clouds have been observed in their interiors. The regions surrounding the elliptical spots are darker than the spots at 41°S. (The GRS will be covered in more detail under a separate heading.) It has been suggested that these elliptical anticyclonic spots are produced by the same process or

Fig. 5.12. Voyager 2 image of the region from 40°N to 40°S taken on June 29, 1979 from 9.3 million km. The white plume cloud is visible (center right) in the EZ to the lower right of the brown spot. The turbulent region (lower right) lies to the west of the Great Red Spot. (Courtesy Jet Propulsion Laboratory/NASA.)

processes that created the GRS. The only real differences would be their physical size and color.

The northern and southern hemispheres consist of similar features but also exhibit a certain degree of asymmetry. In both hemispheres, the elliptical spots display anticyclonic rotation. In the southern hemisphere, elliptical structures such as the GRS, white ovals, and spots at $41^{o}S$ also exhibit a marked degree of similarity. Finally, the EZ is a diffuse cloud region with asymmetrical convection plumes found only near its northern boundary. The hemispheric cloud patterns strongly suggest that strong convection exists on Jupiter.

Convection and Vertical Motion

Voyager imagery provided the proof for the existence of strong vertical convection on Jupiter. Prior to the Voyager photographs, the debate as to what role convection played in Jupiter's global circulation was quite speculative. The resolution levels of earth-based studies revealed the primary zonal circulation but little else. The Pioneer images were better than any earth-based photographs and revealed to a degree, some of the remarkable detail that existed in the cloud patterns. Definite proof of convective activity was still lacking. The Voyager 1 and 2 flyby of Jupiter produced such detailed, high resolution images of convective plumes that little doubt remains concerning the existence of convection on Jupiter.

The plumes appeared as large puffy clouds which extended horizontally for $\approx$2,000 km. Smaller, puffy cumuliform regions 100-200 km in diameter, similar to the tops of earth cumulonimbi, were observed within these plumes. These cumuliform structures have also been observed in all the cloud zones (regions of vertical upward motions) as well as within the GRS. The two cloud colors associated with these convective regions are red and white. Most of the clouds in the convective areas are white. In the GRS, the clouds are red to reddish-orange. A number of suggestions relating to color have previously been mentioned. It is conceivable that the red chromophore of the GRS is derived either from convective activity which is initiated at some different depth in the Jovian atmosphere or related to a phase change in the crystallization process whereby the rate at which phosphine is converted to solid phosphorous determines the color of the cloud tops. The polar regions appeared as mottled regions on the Voyager images. Areas of cyclonic and anticyclonic vorticity as well as what appear to be convective clouds have also been observed in the polar regions. The distribution of convective activity on Jupiter may very well be patterned after the distribution of the energy which causes it. If so, we may speculate that the equatorial convection probably results from the absorption of solar energy whereas the polar convective locations may derive their heat from the planet's own internal heat reservoir. Although such suggestions are conjecture for the present moment, the existence of convective processes has been firmly established as significantly contributing to the morphology of the global cloud pattern.

The Great Red Spot (GRS)

The GRS is the most outstanding individual feature on Jupiter (Fig. 5.13). The Great Red Spot, located at 23°S, was discovered by Cassini in 1665. It has persisted in the Jovian atmosphere for over 300 years. Its latitudinal width (N↔S) is ≈14,000 km and its longitudinal length (W↔E) is ≈25,000 km. Its coloration also varies, most probably in response to the photochemistry of phosphine gas in the Jovian atmosphere. During the later part of the 1960's, the GRS was very faint, whereas during the Pioneer 10 and 11 encounters in 1973 and 1974, the Great Red Spot was very conspicuous as it was for the Voyager 1 and 2 encounters in 1979.

The Great Red Spot is an anticyclonic, counterclockwise, rotating feature of the Jovian southern hemisphere. The clouds associated with the GRS are significantly higher than surrounding clouds and resemble the upper cloud shield generated by a tropical cyclone (hurricane) on earth. Voyager measurements showed that the GRS's clouds tower 15-25 km above the adjacent cloud regions. In fact, several researchers have referred to the GRS and the many smaller spots as Jovian hurricanes. The rotation of the GRS is supported, at least in part, by two oppositely moving currents. On the equatorial side, the wind flow is from the east. On the poleward side, the flow is from the west. The GRS is similar to a disk being spun by two oppositely moving belts. The circulation of the GRS is strongest aloft. The rotation of the atmosphere aloft probably creates an inward flow below the clouds drawing more and more of the atmospheric gases from deeping layers into the vertical column.

A number of theories concerning the origin of the GRS have been suggested, but only two of them appear to be plausible. The first

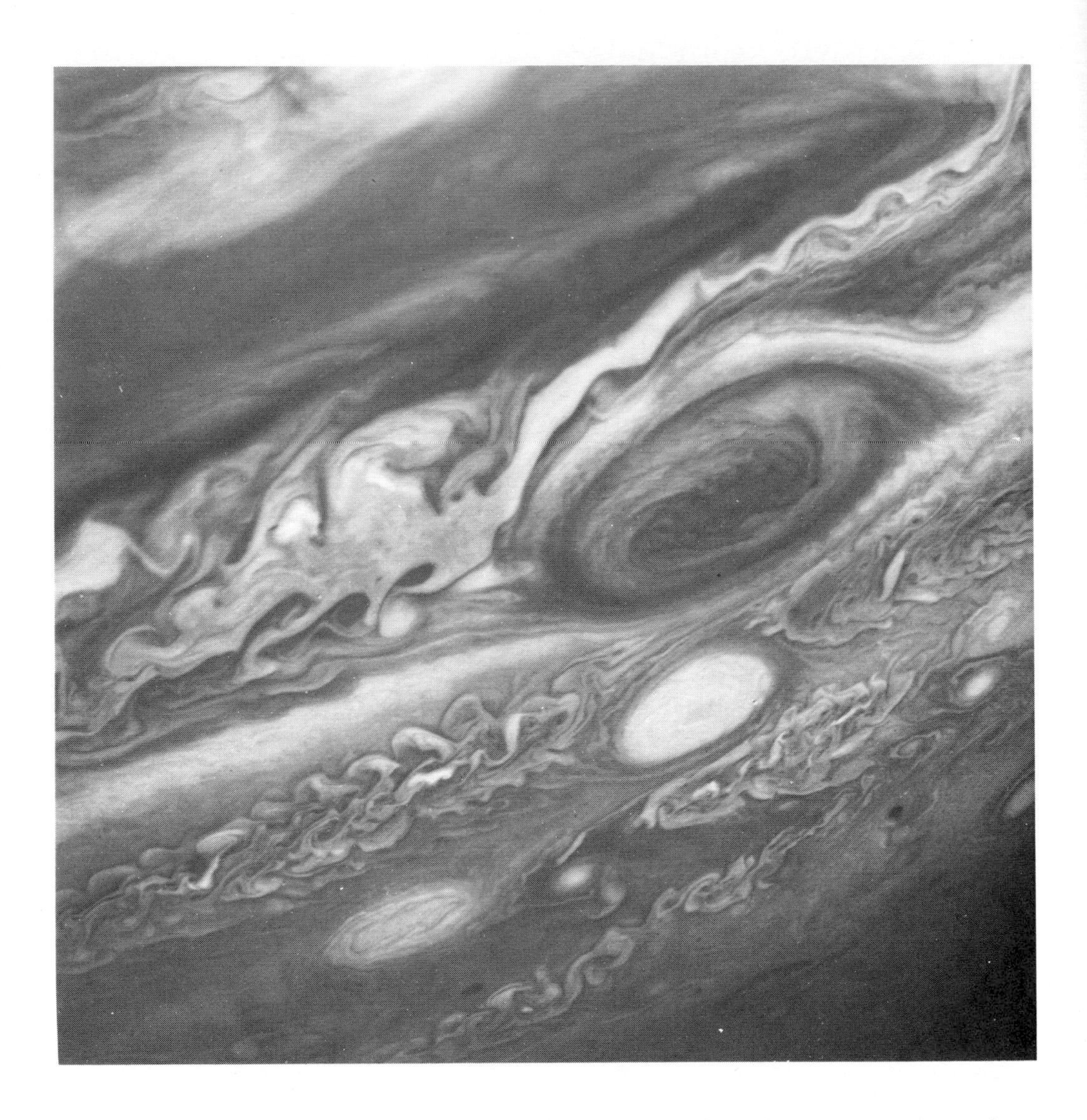

Fig. 5.13. Voyager 2 image shows the region of Jupiter extending from the equator to the south polar latitudes. Image taken on July 3, 1979 from 6 million km. (Courtesy Jet Propulsion Laboratory/NASA.)

and most widely held belief is that the GRS is a form of Jovian storm most closely resembling a tropical cyclone (hurricane) on earth. If the GRS is a Jovian hurricane, then it would explain why it has persisted for over 300 years. Theories which ascribe the existence of the GRS to simply a mass of gas whirling between eastward and westward high speed currents, cannot explain why the red color persists and why mixing and diffusion have not destroyed it altogether. The detailed cloud structure within the GRS suggests strong vertical motions (possibly convective) related to an energy exchange process similar to the release of the latent heat energy in the earth's atmosphere. This, of course, would be the energy sustaining the storm. The longevity of the GRS is probably a function of the radiative time constant. Since the radiative time constant for heat energy dissipation is so long for Jupiter, it would explain how an entity such as the GRS could possibly exist for over three centuries. The disk-like rotation between two oppositely moving jet streams would support and help sustain, but not totally explain, the cause of the GRS.

The second possibility is that the GRS is a solitary wave similar to a crested ocean wave on earth that had not broken (collapsed). Experiments have shown that the crests of solitary waves could rotate.

Occasionally, other spots have been observed interacting with the GRS, sometimes mixing with it, other times traversing the edges of the GRS. These interactions are not mutually exclusive to either of the suggested explanations. The key appears to be the color and the height of the GRS's clouds above the adjacent clouds. These characteristics strongly suggest that the GRS is a massive Jovian storm rather than a solitary wave phenomenon.

The GRS is the only atmospheric circulation of its size and color on Jupiter. There are other entities known as white ovals which display similar anticyclonic circulation. A comparison of the GRS and the white ovals shows that the only differences between them are color and physical dimensions.

In 1879, the GRS measured 40,000 km (E-W) by 13,000 km (N-S). By 1979, its length had shrunk to 25,000 km while its width remained at 13,000 km. It is obvious that the GRS is shrinking in a longitudinal (E-W) direction. Its rate of shrinkage and drift rate around the planet have remained relatively linear over the past 40 years.

Three white ovals, each 26,000 km in length (E-W), formed following a disturbance in the STeB in 1938. (A typical white oval is located just south (beneath) the GRS in Fig. 5.13.) Their present length has decreased by 50 percent to ≈12,000 km. Their drift rate in 1941 was 7.5 m sec^{-1}. It is now 4.0 m sec^{-1}. The drift rate and shrinkage rate of the white ovals decreased dramatically from 1943 to 1966. Since 1967, these changes have been more linear.

Estimates show the GRS will shrink to 24,500 km by 1985 while the white ovals will decrease from their present 12,000 km to 9,700 km. Perhaps these changes indicate that the GRS is a Jovian storm that is slowly degenerating. More detailed analyses of these changes will eventually provide the clues necessary to an understanding of these phenomena.

The sequential Pioneer 10 and 11 missions followed by Voyager 1 and 2 permitted an analysis of detailed temporal and spatial changes in the Jovian cloud patterns. The 12 months that elapsed between Pioneer 10 and 11 produced very few changes in the cloud structure of Jupiter. Since the Pioneer equipment did not have the resolving power of the Voyager series, it is possible that changes occurred but the equipment was not sufficiently sensitive enough to resolve these changes.

During the interval between Pioneer 11 and Voyager 1 (4.25 years), a disturbance of the SEB along the northern margin of the GRS darkened and the region between 23°-35°N brightened. The changes that occurred in the four months between Voyager 1 and 2 were literally unbelievable. The rapidity of these changes in such a short time span has prompted scientists to reconsider their temporal framework for Jupiter's atmosphere. Prior to Voyager, Jupiter's planctary atmospheric featurcs were considered quasi-permanent. Significant changes were expected to take four to five years to occur. The magnitude and number of changes in the intervening four months between probes revealed a dynamic atmosphere where almost anything could and did happen.

Several noticable changes near the GRS were observed (Fig. 5.14). The white oval in Fig. 5.14 (feature 1) moved eastward from 80°W to

Fig. 4.11. Frost filled Argyre Basin just before dawn. Note water-ice plume cloud on the western flank of Ascreaus Mons. (Courtesy Jet Propulsion Laboratory/NASA.)

Fig. 5.1. Voyager 1 photograph of Jupiter taken on January 29, 1979 from 35.6 million km. (Courtesy Jet Propulsion Laboratory/NASA.)

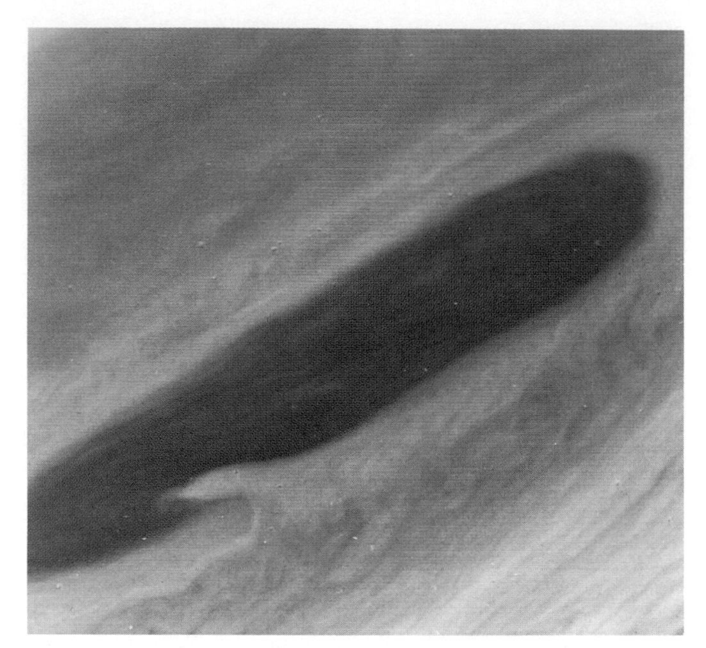

Fig. 5.11. Voyager 2 image of one of the long, dark clouds observed in the NEB taken on July 6, 1979 from 3.2 million km. Thin white clouds are also seen within the dark cloud. (Courtesy Jet Propulsion Laboratory/NASA.)

Fig. 5.13. Voyager 2 image shows the region of Jupiter extending from the equator to the south polar latitudes. Image taken on July 3, 1979 from 6 million km. (Courtesy Jet Propulsion Laboratory/NASA.)

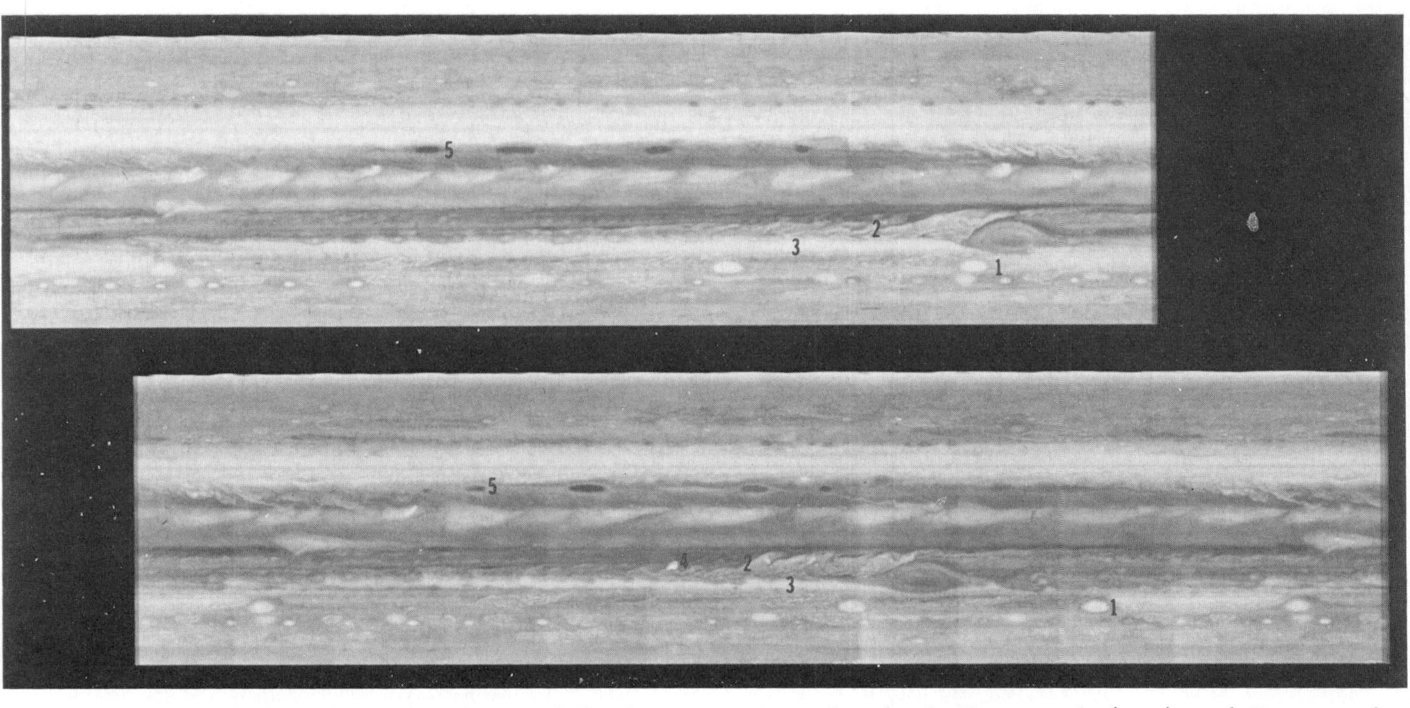

Fig. 5.14. Cylindrical projections of Jupiter representing both Voyager 1 (top) and Voyager 2 (bottom). The longitude scale is correct for both images with the right edge 0 degrees. The comparison shows the relative motions of features in Jupiter's atmosphere. (Courtesy Jet Propulsion Laboratory/NASA.)

40°W. The longitudinal extent of the cyclonic area west of the GRS expanded (Fig. 5.14, feature 2). The STeZ's width (Fig. 5.14, feature 3) has shrunk. An intensely bright, white cloud (Fig. 5.15, feature 4) appeared while velocities of the current on the southern edge of the EZ dropped from 150 to 95 m sec^{-1}. The high velocity equatorial current (100 m sec^{-1}) moved latitudinally 2° closer to the equator. A dark brown ellipse (Fig. 5.14, feature 5), shrunk and appears to be disappearing. Voyager 1 detected 13 plumes along the northern edge of the EZ. Voyager 2 found only two plumes. The different positions of the features labeled on Fig. 5.14 confirm the dramatic changes that have occurred in four short months (February-May, 1979) between Voyager 1 and 2. This added complexity of the Jovian atmosphere is now a fact that scientists must deal with in their attempts to model the planet's general circulation.

Cloud Classification

The cloud data from the Voyager probes prompted Ingersoll et al. (1979) to attempt a classification of Jupiter's cloud patterns based on cloud texture. Texture was chosen as the basis for the classification because it provides insight into the dynamic processes which create and shape the visible clouds. There are nine categories derived from the various cloud patterns seen in the images returned by Voyager 1 and 2. The nine categories are summarized in Table 5.2 and correlated to actual photographs and inset sketches (Fig. 5.15).

A map of the Jovian clouds based on this classification is presented in Fig. 5.16. The equatorial zone exhibits the least textural contrast, the only exception being the plume heads. The remainder of the planet continues to reveal banded patterns of cloud types. The value of the textural classification is that it provides a useful tool for cataloging cloud types and a means of recording morphological changes that might occur within each cloud region. Observed changes in the cloud patterns may provide more clues to the processes at work deeper in the atmosphere below the cloud tops.

Radio Emissions from Jupiter

Decametric radio emissions from Jupiter were first detected in 1955 and have been closely studied since that time. By 1964, it was readily apparent that Jupiter's moon, Io, was responsible for modulating the decametric emissions from Jupiter. Io's orbit lies well within Jupiter's magnetosphere and strongly interacts with it

TABLE 5.2

A classification of Jovian clouds by texture (after Ingersoll et al., Nature, 280, p774, 1979. Copyright 1979 Macmillin Journals Ltd.)

Type	Description
a	round puffy clouds with irregular organization (suggestive of small scale cumulus on earth)
b	puffy cumuliform clouds in linear array (suggestive of convection with lateral shear zones)
c	linear cloud pattern (suggestive of strong shear and/or waves with irregular variations along the wave crests)
d	an isolated bright region with puffy clouds along the edge
e	filaments with long parallel threads (suggestive of laminar flow patterns)
f	a filamentary spiral opening outwards in either direction
g	folded filaments or ribbon-like pattern
h	a V-shaped pattern
i	a region of low texture

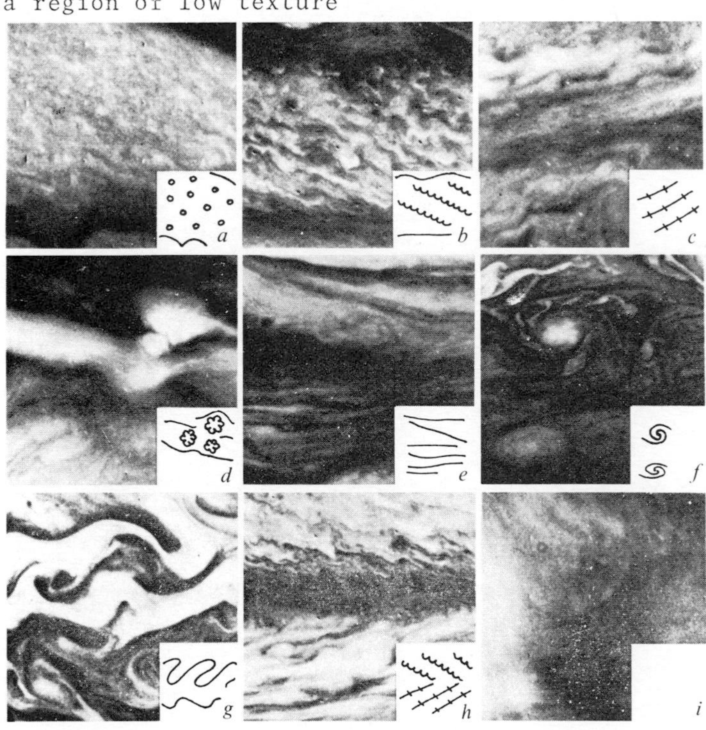

Fig. 5.15. Cloud classification based on textural types (after Ingersoll et al., Nature, v 280, p. 790, 30 Aug. 1979. Copyright 1979 by the Macmillin Journals Ltd.)

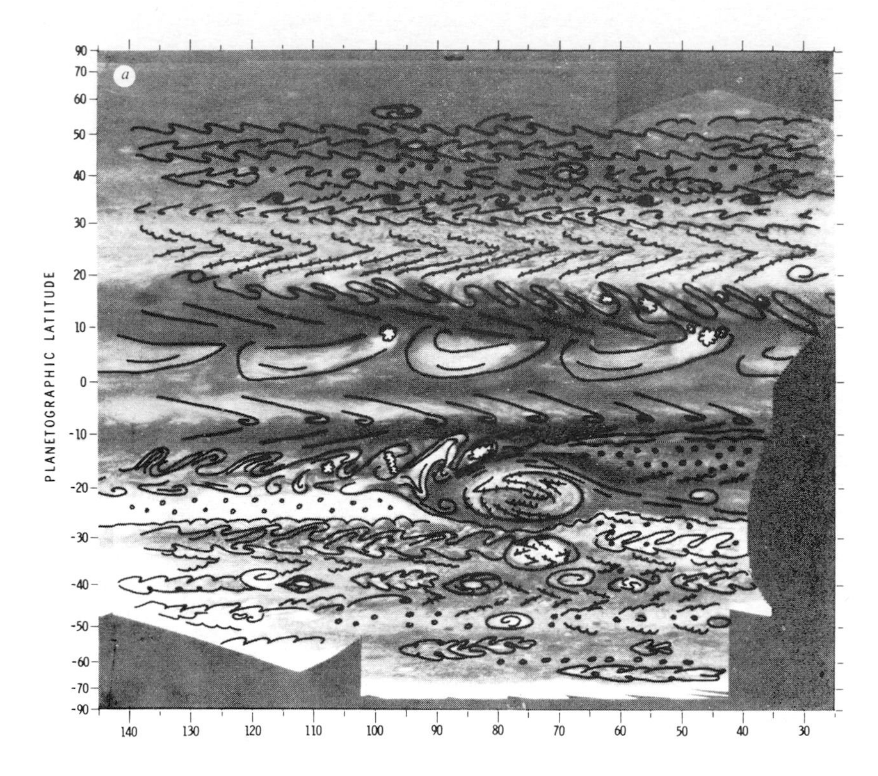

Fig. 5.16. Textural map of Jupiter's cloud types based on the classification by A.P. Ingersoll et al.. (Courtesy A.P. Ingersoll et al., Nature, v 280, p. 791, 30 Aug. 1979. Copyright 1979 by the Macmillin Journals Ltd.)

producing the plasma torus and flux tube phenomena discussed earlier in this chapter. The decametric radio emissions are generated along the magnetic field lines that connect Jupiter and Io and also conduct a 400,000 volt, 5,000,000 ampere current. The primary source region for these radio emissions appears to be located 20° in advance of Io's orbital position, where the foot of Io's flux tube touches the Jovian ionosphere.

These radio emissions are generated when Io is positioned within the northern magnetic hemisphere. The emissions are even stronger if the Jovian north magnetic pole happens to be tilted toward Io. The radio frequencies match the electron gyrofrequencies at the flux tube foot. The radio waves thus emitted form a cone or cylinder-like emission pattern. Earth-based receivers are only able to monitor these decametric emissions sporadically when an appropriate receiving geometry is present. Several conditions must be met for an appropriate viewing geometry. First, Io must be located in the northern magnetic hemisphere. Second, Io must not be occulted by Jupiter. Third, the orientation of the Earth - Io/Jupiter geometry must be such that the radio emissions do not pass through and hence are not refracted by Jupiter's atmosphere. The generation of radio emissions through the interaction of Io, and Jupiter's magnetosphere and ionosphere represent one of the most unique processes discovered in the study of planetary atmospheres.

Recently, non-Io related decametric emissions from Jupiter have been monitored. These emissions do not originate from a specific area but seem to radiate from across the planet's disk. It appears that approximately ten days after a period of intense solar activity, the enhanced solar wind reaches, compresses, and energizes the Jovian magnetosphere. The result is more decametric radio emissions. These emissions have been labeled as non-Io related activity (NIR).

Voyager 1 detected a different type of radio emission from Jupiter in October, 1978 when the spacecraft was still ≃155,000,000 km from Jupiter. The energy frequencies ranged from 10-56 kHz and were also detected by the Voyager 2 probe. These kilometric emissions were storm-like with typical bursts lasting approximately one hour. The duration of the bursts varied greatly. Some emissions lasted only a few minutes while others persisted for more than four hours. The emission characteristics varied with the duration of the outburst. In general, the frequency of the emissions decreased with time. The kilometric emissions appeared to originate from two source regions: one at ≃200°W and a less powerful center near 20°W. Three causes have been suggested for the kilometric radio emissions: electrical activity within Io's torus, auroral displays, and the volcanic activity on Io's surface. Electrical activity often accompanies major volcanic eruptions on earth so it is logical to assume it may also accompany Io's volcanoes. If electric arcs are associated with Io's volcanoes, they may be a source for some, but not all, of the kilometric radio emissions since the radio waves seem to come from several sites near both Io and Jupiter. The volcanic electric arc theory would only explain the emissions from the region near Io. These emissions would be as periodic as the volcanic activity on Io. It is therefore highly unlikely that the volcanoes are the primary source of electrical activity and kilometric radio emission. The evidence points more to the plasma torus and/or auroral displays, however, the answer to this puzzle may have to wait until the Galileo spacecraft reaches Jupiter in 1985.

Aurora

Auroral displays have been observed in three zones, 700 km, 1,400 km, and 2,300 km above Jupiter's cloud surfaces. The auroral zone is restricted to $\simeq 10^{\circ}$ of latitude from the magnetic pole. The observed auroral arcs were over 30,000 km long, of very short duration (<1 min.) and very bright. The kilometric radio emissions from the auroral regions are similar to those recorded during earth auroral displays.

Lightning

Whistlers and direct observations have confirmed the existence of lightning on Jupiter. Whistlers are low frequency electromagnetic waves produced by lightning and propagated along the magnetic field lines. The noise that they produce on a radio receiver sounds just like a whistle, hence the name whistlers. The existence of whistlers is direct proof of lightning discharges.

Multiple lightning images have been photographed on the dark side of Jupiter by both Voyager 1 and 2 (Fig. 5.17). Discrepancies exist in the reports of lightning distribution. One source reports a uniform distribution of lightning across the planet while another states that the discharges are clustered near 30° and 46°N latitude. This issue is, as yet, unresolved. The lightning discharges have been described as regions of bright arcs surrounded by a glow. One observation reported that a convective cloud appeared to be internally illuminated by a lightning discharge.

Lightning bolts on Jupiter are typically more powerful than the terrestrial lightning bolts particularly those bolts associated with the foot of Io's flux tube. The electrical discharges observed in association with these cumuliform-like clouds suggest that these regions may experience a charge segregation similar to that established between the lower portions of cumulonimbi and the ground

Fig. 5.17. Voyager 1 image of lightning on Jupiter's dark side. This 3 minute, 12 second exposure was teken on March 5, 1979. Jupiter's north pole is top center. The double bright streak represents auroral activity above the north polar regions. (Courtesy Jet Propulsion Laboratory/NASA.)

on earth. If so, it may be possible to monitor and map the global distribution of convective activity on Jupiter through the identification of regions with lightning activity.

Jupiter's Ring

Voyager 2 photographed a Jovian ring different from any observed to date (Fig. 5.18). Analysis of the ring's characteristics revealed its inner boundary at ≃1.68 R_J and its outer boundary at ≃1.8 R_J. The inner edge consists of a very faint region from 1.68 to 1.7 R_J. The region beginning at 1.7 R_J is brighter and extends for ≃800 km where it grades into a faint outer edge region extending to 1.8 R_J. The upper limit for the ring's thickness has been set at ≃30 km and the mean diameter of the particles making up the ring at 4 μm.

Photographs of the ring show that the ring's outer edge seems to terminate more abruptly than the inner edge. The newly discovered moon at 1.81 R_J may account for the ring's abrupt end by a sweeping action whereby the particles in the orbital path are collected by the moon's weak gravitational attraction. The inner edge of the ring may not actually terminate at 1.68 R_J. A faint glow has been detected between 1.68 R_J and the ionosphere, strongly suggesting that the interior of the ring stretches down to the atmosphere. The presence of ring particles might also account for the heretofore unidentified absorber of ultraviolet energy high above the planet's cloud surfaces, but only if the high level stratospheric winds distribute these particles planetwide.

The discovery of this ring raised the issue of the nature of the ring and the source of the material present. It is unlikely that this ring was formed during the condensation events that formed the solar system, as perhaps Saturn's rings were. The ring does not appear to be the remnants of a former Jovian satellite. The ring appears to be the result of a steady-state process whereby particles are gained and lost. Losses occur to both the newly discovered moon at 1.81 R_J and the Jovian gravitational and magnetic fields. Particles gained are either sputtered (ejected by the collision of high energy particles with Io's surface) or volcanically ejected from Io, the innermost Jovian moon. The ring is therefore dynamically related to both Jupiter and Io.

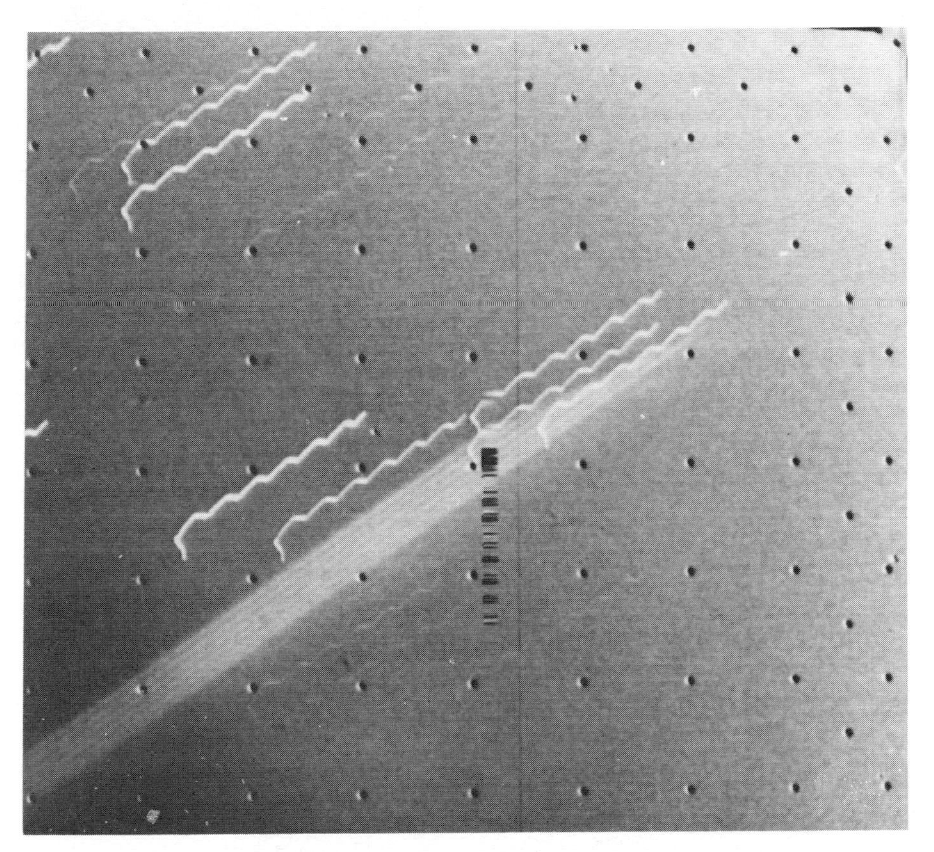

Fig. 5.18. Discovery photo (11 minute exposure) of Jupiter's ring taken by Voyager 1 on March 4, 1979. The multiple exposure of the extremely faint ring appears as a broad light band crossing the center of the picture. The background stars look like hair pins because of spacecraft motion during the time exposure. Ring thickness is ≃30 km. (Courtesy Jet Propulsion Laboratory/NASA.)

Summary

Jupiter is still a planet with many unanswered questions. It has an atmosphere in which the relative ratios of constituents are more like the sun then any of the other planets. Yet, there are numerous organic compounds suggesting a primordial atmosphere similar to what scientists suspect may have existed on earth some three to four billion years ago. Its (Jupiter's) energy budget is equally mystifying. The planet radiates 1.9 times more energy than it receives from the sun belt but in a most unusual pattern. The bulk of the absorbed solar energy is reradiated to space from the tropical and subtropical latitudes while energy from an as yet unexplained internal source is radiated out to space from the polar regions. Is, or was, Jupiter a proto-star that did not have sufficient mass to evolve any further? Many of its characterists certainly point in that direction.

Jupiter's atmospheric mass of hydrogen and helium has often been compared to the earth's oceanic circulation whose upper portions are vigorous and dynamic while in the lower regions, flows are generally sluggish and weak. In the oceans, these sluggish, deep water circulations result from the extreme stability created by the water's uniform temperature and salinity characteristics. It would seem that it is at this point that the analogy between Jupiter's atmosphere and the earth's oceans dissolves. The lower Jovian atmosphere is far from stable. It is heated by an, as yet, unexplained internal heat source as well as by solar radiation. Convective motions are generated, and supported by a near adiabatic lapse rate in the troposphere. The mechanisms that produce this absorption of incoming solar energy are not fully understood. Several explanations were offered in the opening sections which covered the composition and structure of the Jovian atmosphere.

The cloud bands and winds indicate that strong zonal circulations dominate the tropical and subtropical latitudes but poleward of 40°-50°, the banded pattern is replaced by a mottled pattern of relatively small, turbulent eddies as well as larger, circular phenomena (1,600 km in diameter) which may be comparable to cyclonic vortices here on earth. Convective phenomena, lightning, and aurora also occur in the polar regions.

The Great Red Spot and other smaller anticyclonic ovals suggest patterns similar to tropical cyclones on the earth, however, the analogy must end there for the Great Red Spot is over 300 years old. Perhaps the long radiative time constant for Jupiter explains how individual atmospheric phenomena can last for over three centuries.

Whether or not Jupiter has a solid or liquid core is uncertain. The evidence seems to suggest a small, rocky inner core surrounded by liquid metallic hydrogen. Jupiter does have a strong magnetic field and magnetosphere. The collapsing of the Jovian magnetosphere during periods of solar wind intensification is one of the most unusual phenomena observed by the Voyager probes. The plasma is compressed and plasma electrons are accelerated to velocities approaching the speed of light. These electrons are ejected from Jupiter's orbit. Some reach the vicinity of the earth every 13 months when Jupiter and the earth are connected by the lines of the interplanetary magnetic field, explaining the heretofore, anomalous increase in high-speed electrons observed by earth orbiting satellites every 13 months.

The interactions between Io and Jupiter's magnetosphere are worthy of a separate volume. The highly energetic plasma particles, a plasma torus of sodium, oxygen, and sulfur ions, a 400,000 volt, 5,000,000 ampere flux tube current, lightning, auroral displays, and even the newly discovered ring may all be attributed to, or somehow related to, Io/Jupiter interactions.

The Pioneer and Voyager probes enabled us to learn more about Jupiter and its moons over a span of six years (1973-79) than man had learned in over 300 years of telescopic observations. Without a doubt, Jupiter's atmospheric processes rival Venus' in uniqueness and complexity. Our study of Jupiter's atmosphere warrants very special attention for it may represent a mirror image of how the earth's atmosphere looked eons ago at the birth of our solar system.

References

Beatty, J.K., The far-out worlds of Voyager I - II, Sky & Telescope, 57, 516 ff., 1979.

Beebe, R.F. et al., Pre-Voyager velocities, accelerations, and shrinkage rates of jovian cloud features, Nature, 280,771 ff., 1979.

Bridge, H.S. et al., Plasma observations near Jupiter: initial results from Voyager 1, Science, 204, 987 ff., 1979.

Bridge, H.S. et al., Plasma observations near Jupiter: initial results from Voyager 2, Science, 206, 972 ff., 1979.

Broadfoot, A.L. et al., Extreme ultraviolet observations from Voyager 1 encounter with Jupiter, Science, 204, 979 ff., 1979.

Chase, S.C. et al., Pioneer 10 IR radiometer experiment: preliminary results, Science, 183, 315 ff., 1974.

Cook, A.F. et al., A lower limit on the top of Jupiter's haze layer, Nature, 280, 780 ff., 1979.

Cook, A.F. et al., First results on jovian lightning, Nature, 280, 794 ff., 1979.

Eshleman, V.R. et al., Radio science with Voyager 1 at Jupiter: preliminary profiles of the atmosphere and ionosphere, Science, 204, 976 ff., 1979.

Eshleman, V.R., Radio science with Voyager at Jupiter: initial Voyager 2 results and a Voyager 1 measure of the Io torus, Science, 206, 959 ff., 1979.

Gurnett, D.A. et al., Whistlers observed by Voyager 1: detection of lightning on Jupiter, Geophys. Res. Let., 6, 511 ff., 1979.

Hanel, R. et al., Infrared observations of the jovian system from Voyager 1, Science, 204, 972 ff., 1979.

Hanel, R. et al., Infrared observations of the jovian system from Voyager 2, Science, 206, 952 ff., 1979.

Hord, C.N. et al., Photometric observations of Jupiter at 2400 angstroms, Science, 206, 956 ff., 1979.

Hubbard, W.B. and Jokipii, J.R., New studies of Jupiter, Sky & Telescope, 50, 212 ff., 1975.

Ingersoll, A.P., The meteorology of Jupiter, Scientific American, 234, 46 ff., March 1976.

Ingersoll, A.P. et al., Zonal velocity and texture in the jovian atmosphere inferred from Voyager images, Nature, 280, 773 ff., 1979.

Kivelson, M.G. et al., Magnetospheres of the galilean satellites, Science, 205, 491 ff., 1979.

Kliore, A. et al., Preliminary results on the atmospheres of Io and Jupiter from the Pioneer 10 - s band occultation experiment, Science, 183, 323 ff., 1974.

Krimigis, S.M. et al., Low energy charged particle environment at Jupiter: a first look, Science, 204, 998 ff., 1979.

Krimigis, S.M. et al., Hot plasma environment at Jupiter, Science, 206, 977 ff., 1979.

Kurth, W.S., Low frequency radio emissions from Jupiter: jovian kilometric radiation, Geophys. Res. Let., 6, 747 ff., 1979.

Lucadamo, F.J. and Panofsky, H.A., On a relation between winds on the earth and Jupiter, Bull. American Meteor, Soc., 59, 700 ff., 1978.

Mitchell, J.L. et al., Jovian cloud structure and velocity fields, Nature, 280, 776 ff., 1979.

Ness, N.F. et al., Magnetic field studies at Jupiter by Voyager 1: preliminary results, Science, 204, 982 ff., 1979.

Ness, N.F. et al., Magnetic field studies at Jupiter by Voyager 2: preliminary results, Science, 206, 966 ff., 1979.

Ness, N.F. et al., Jupiter's magnetic tail, Nature, 280, 799 ff., 1979.

Owen, T. et al., Jupiter's rings, Nature, 281, 442 ff., 1979.

Pilcher, C.B., Images of Jupiter's sulfur ring, Science, 207, 181 ff., 1980.

Prentice, A.J.R. and ter Haar, D., Origin of the jovian ring and the galilean satellites, Nature, 280, 300 ff., 1979.

Prinn, R.A. and Lewis, J.S., Phosphine on Jupiter and implications for the Great Red Spot, Science, 190, 274 ff., 1975.

Sandel, B.R. et al., Extreme ultraviolet observations from Voyager 2 encounter with Jupiter, Science, 206, 962 ff., 1979.

Smith, E.J. et al., Jupiter's magnetic field, magnetosphere, and interaction with the solar wind: Pioneer II, Science, 188, 451 ff., 1975.

Smith, B.A. et al., The Jupiter system through the eyes of Voyager 1, Science, 204, 951 ff., 1979.

Smith, B.A. et al., The galilean satellites and Jupiter: Voyager 2 imaging science results, Science, 206, 927 ff., 1979.

Stone, E.C. and Lane, A.L., Voyager 1 encounter with the jovian system, Science, 204, 945 ff., 1979.

Stone, E.C. and Lane, A.L., Voyager 2 encounters with the jovian system, Science, 206, 925 ff., 1979.

Soderblom, L.A., The galilean moons of Jupiter, Scientific American, 242, 88 ff., 1980.

Terasawa, T. et al., Solar wind effect on Jupiter's non Io related radio emission, Nature, 273, 131 ff., 1978.

Terrile, R.J. et al., Infrared images of Jupiter at 5-micrometer wavelength during the Voyager 1 encounter, Science, 204, 1007 ff., 1979.

Terrile, R.J. et al., Jupiter's cloud distribution between the Voyager 1 and 2 encounters: results from 5 micrometer imaging, Science, 206, 995 ff., 1979.

Vogt, R.E. et al., Voyager 1: energetic ions and electrons in the jovian magnetosphere, Science, 204, 1003 ff., 1979.

Vogt, R.E. et al., Voyager 2: energetic ions and electrons in the jovian magnetosphere, Science, 206, 984 ff., 1979.

Warwick, J.W., Voyager 1 planetary radio astronomy observations near Jupiter, Science, 204, 995 ff., 1979.

Warwick, J.W. et al., Planetary radio astronomy observations from Voyager 2 near Jupiter, Science, 206, 991 ff., 1979.

Wolfe, J., Jupiter, Scientific American, 233, 118 ff., 1975.

CHAPTER VI

IO AND GANYMEDE: MOONS OF JUPITER

IO

Descriptive Statistics:

Diameter	3,640 km
Distance from Jupiter	352,000 km
Mean Density	3.5 g cm^{-3}
Period of Rotation	none (always presents same face to Jupiter)

Introduction

Io is neither the largest natural satellite in our solar system nor is it the smallest. It is, however, one of the most interesting bodies because of its magnetic and gravitational interactions with Jupiter.

Io's Volcanoes

A discussion of Io's volcanism must, of necessity, precede those sections of the chapter concerned with atmospheric processes. Io's active volcanoes were fortuitously discovered by L. Morabito of the Jet Propulsion Laboratory. Her discovery, combined with the work of others, provided answers to questions about Io that have puzzled scientists for several decades.

Voyager 1 detected eight volcanoes on Io. Voyager 2 observed seven of the eight volcanoes on its flyby and reported that six were still actively erupting. Various theories have been suggested to explain the heating necessary to create the volcanic eruptions on Io. The two most probable explanations are the electrical (ohmic) heating and the tidal heating theories. The electrical heating theory views Io as a large electrical resistor. The 400,000 volt, 5,000,000 ampere current in the flux tube is conducted through Io's ionosphere and planetary mass. Regions of low resistance permit the current to flow more easily. These low resistance areas may be the volcanic calderas. Conceivably, electric arcs are struck, much the same as with an arc welder, vaporizing surface materials into a column of hot expanding gases. The column of gases expands upward cooling rapidly and condensing into solid particles, which produce the ballistic plumes observed in the Voyager images (Gold, 1980).

A variation of the arc/resistance theory eliminates the need for a surface arc. Instead, the current passes through Io's interior and heats subsurface reservoirs of sulfurous materials. The subsurface pressures increase, finally erupting upward as a volcanic plume.

Proponents of the tidal heating explanation suggest that Io must flex, like a rubber ball, in response to Jupiter's tides. The tide-induced flexing produces heat, much like the heat generated by flexing a piece of metal wire. The heat would melt the subsurface materials resulting in their liquification. The increased pressures would eventually produce a volcanic eruption. It may be that a composite theory combining both ideas may eventually be proven correct but the end result is the same. Io is a moon disfigured and modified by volcanic activity and best described as looking like a "pizza with small pox" (Fig. 6.1).

Plume characteristics vary from one volcano to another on Io. Plume 1, at the center of the heart shaped region in Fig. 6.1 was ≃35 km wide at the base of its fountain and extended vertically to ≃280 km above Io. The material being ejected is probably sulfur or sulfur dioxide (Fig. 6.2).

Atmospheric Composition

Sulfur dioxide has been positively identified as the primary constituent of Io's atmosphere. There is also strong evidence pointing to the deposition of solid SO_2 on Io's surface. The existence of an SO_2 atmosphere on Io certainly helps to explain the source of the sulfur, oxygen, and possibly sulfur dioxide ions in the plasma torus. The SO_2 lost from Io's tenuous atmosphere undergoes ionization and becomes incorporated into Io's torus.

Atmospheric Pressure

Io's atmosphere is so thin that surface pressure is estimated at only 0.1 microbar or one ten millionth of the earth's surface pressure. The erupting volcanoes explosively eject and propel fluid sulfur dioxide vertically upward in a ballistic trajectory. The SO_2 freezes into sulfur dioxide frost crystals, the bulk of which fall

Fig. 6.1. Io taken by Voyager 1 on March 4, 1979 from 862,000 km. The doughnut-shaped feature in the center is the site of an erupting volcano. (Courtesy Jet Propulsion Laboratory NASA.)

Fig. 6.2. Voyager 1 image of erupting volcano on Io's limb taken on March 4, 1979. The distance to Io was 490,000 km. The plume height is ≈180 km. (Courtesy Jet Propulsion Laboratory/NASA.)

back toward Io's surface for recycling. Some of the sulfur dioxide escapes into the torus.

Io's colors may result from different forms of sulfur in the sulfur dioxide. The color of sulfur is temperature dependent, ranging from yellow to black. If the temperature of the sulfur is suddenly lowered, as it would be when ejected in a volcanic plume, the color remains the same. Io's surface coloration may therefore result from flash frozen sulfur dioxide compounds which originated at volcanic vents of different temperatures. Recent data (Nelson et al., 1980) based on the spectral responses of sulfur dioxide and alkali sulfides show that the white regions on Io are SO_2 frost and the red regions are primarily other forms of sulfur. This should not be construed to indicate that SO_2 is absent in the red regions but rather that its spectral response suggests that it is relatively less abundant than the other sulfur compounds. Their (Nelson et al.) results also help to explain the changes in the patterns of white and red regions observed by Voyager 1 and 2 on their flyby of Io. Since the volcanic eruptions are episodic, changes in the surface coloration of Io should occur both spatially and temporally.

Io's Ionosphere

Pioneer 10's occultation data revealed an ionosphere which extends ≈700 km above Io's surface. Other published data suggest a very thin ionosphere 20,000 times less dense than the earth's ionosphere and only 110 km deep. Voyager data also places the height of the day side ionosphere at 700 km. The night side ionosphere is much shallower and subject to sweeping by the strong Jovian magnetic field. An electron density peak of 6×10^{-4} electrons cm^{-3} was also reported.

Io's ionosphere is unique because it lies within the boundaries of Jupiter's magnetosphere. Io's motion through Jupiter's magnetosphere creates a distribution of electrical charges. As previously stated, when Io reaches a certain position with respect to Jupiter's magnetic field, a current is generated. This current follows a circuit between Io's ionosphere and Jupiter's ionosphere and back again along the lines of Jupiter's magnetic field. This burst of energetic electrons generates the dekametric radio signals associated with Jupiter.

Temperature Structure

Surface temperature measurements indicate that the side of Io facing Jupiter experiences temperatures of about $130^{o}K$ while temperatures near the volcanic vents may range from 290°-$400^{o}K$. The plasma torus temperatures are much higher temperatures reaching $10^{5o}K$.

If the electric arc theory of volcanic eruptions is correct, the temperatures at Io's surface in the volcanic calderas where the arc strikes should reach $6,000^{o}K$. This value is based on the temperature needed to vaporize sulfur compounds and propel them upward at an escape velocity of 1 km sec^{-1}.

Albedo

Io's albedo is 0.17 but surface brightness varies temporally and spatially. Telescopic observations have noted that Io was significantly brighter as it entered sunlight after having been in Jupiter's shadow for 21 hours. Its albedo increased from 0.17 to 0.61 suggesting a condensable atmosphere. Various explanations for these post eclipse brightenings (PEB's) have been offered. Historically, these PEB's were thought to represent some form of snow on Io and Ganymede. A methane (CH_4) snow seemed to be the most probable explanation at that time. Others had suggested that Io's brilliance is due to evaporite minerals or "salts" of sodium and potassium deposited as water was outgassed and evaporated. This explanation was at least plausible, in light of evidence which shows that Io is as reflective in the infrared spectrum as it is in the visible spectrum. This characteristic is shared with few materials, among them the Bonneville Salt Flats of Utah. Voyager 2 data confirmed that a whitish SO_2 frost may exist on Io, particularly at the poles and on Io's dark side where temperatures are lower. It is possible that when Io goes behind Jupiter, temperatures in the normally sunlit atmosphere plummet and an SO_2 frost condenses out on the surface. The condensate evaporates within ten minutes as Io enters sunlight. But for those ten minutes, Io is the most reflective object in our solar system!

Another theory relates Io's post eclipse brightenings to solar flares. Energetic particles emitted by solar flares may bombard Io's surface producing something akin to an astronomical thermoluminescence. The energy absorbed may be emitted as visible light in response to reactions triggered by Io's return into sunlight. Again, the entire process ends within a time span of ten minutes. For the present, the sulfur dioxide frost seems to be the most plausible explanation but it does not satisfactorily explain the erratic occurrences of post eclipse brightenings.

The erratic PEB's might be explained by the concentration of SO_2 in Io's atmosphere. The amount of SO_2 must vary as Io loses some of its atmosphere to the torus. It must also change in response to the number of active volcanic plumes. Not all the volcanoes observed by Voyager 1 were erupting as Voyager 2 photographed Io, therefore, temporal variations in volcanic activity do exist. The concentration of SO_2 in Io's atmosphere at any given moment is a function of the rate at which it is supplied to the atmosphere by the volcanoes. What if a critical lower limit of SO_2 is reached as volcanic activity slows down or ceases? Obviously, the formation of the SO_2 frost would either cease or not be as noticable. The result would be erratic occurrences of PEB of Io in association with temporal variations in volcanic activity. Perhaps the Galileo mission in 1985 will provide the answers to this interesting problem.

Magnetosphere

The Io plasma torus' interactions with Jupiter have already been discussed but several unanswered questions remain. A review of the torus' characteristics would be beneficial before proceeding. The torus is centered at ≈6 R_J and is ≈1 R_J thick. It lies within the equatorial plane of Jupiter's magnetodisk and oscillates up and down in response to Io's orbital variations. The primary constituents of the plasma are sodium, sulfur, oxygen, and possibly sulfur dioxide ions. The thermal electron temperature of the plasma is 10^{5} °K with temperatures decreasing rapidly Jupiterward. The Voyager flyby passed close enough to the torus to note asymmetrical variations in electron concentrations. The number of electrons averaged 1,000 cm^3. Voyager 2 was in the outer periphery of the torus for about three hours. It observed an electron concentration increase, a slight dip, then another peak, and then a decrease as it left the outer torus. The reasons for these variations are not yet clear.

The source for the materials in the torus is Io. SO_2 is volcanically ejected and ionized. The neutral sodium is sputtered off Io and accelerated to velocities of ≈55 km sec^{-1} by the impact of particle collisions with the surface. The distribution of sodium reveals that there is more sodium in the torus ahead of Io. We would normally expect to find more sodium in the torus behind Io. The sodium cloud is also skewed inward toward Jupiter. Io's ionosphere is another problem. How is it that Io, with such a weak gravitational field, can maintain an ionosphere against the pressure of the faster corotating Jovian plasma? The existence of both the asymmetrical sodium distribution and the ionosphere could be explained if Io has a magnetic field and magnetosphere of its own. The magnetosphere would shield Io from the Jovian plasma. The sodium sputtered off in front of Io would be shielded for a longer period of time by the magnetic field. The field would also interact with Jupiter's magnetic field, skewing it and the sodium toward Jupiter or in a direction predeeding Io. The existence of interacting magnetospheres could also explain why the location of Io's flux tube's foot above Jupiter is 20° ahead of Io. The evidence is, as yet, circumstantial but it certainly strongly suggests that Io may have its own magnetosphere.

Summary

Io is the closest Jovian satellite and traditionally has been relegated to a secondary status in studies of the solar system. However, the combination of volcanism, an unusual, condensable atmosphere, large albedo variations, radio emissions, and ionospheric interactions with the Jovian magnetospheric flux make Io one of the most fascinating and dynamic entities in the solar system.

GANYMEDE

Descriptive Statistics:

Diameter	5,270 km
Distance from Jupiter	1,000,000 km
Mean Density	2.03 g cm^{-3}
Period of Rotation	none (always presents same face to Jupiter)

Atmosphere

Evidence for an atmosphere on Ganymede, the largest of Jupiter's satellites, was obtained in 1973. Astronomical observation of Ganymede's occultation of the star, SAO 186800 in 1972, revealed a gradual extinction of the star's light as the moon's limb passed between the star and earth-based observatories. This gradual extinction was interpreted as direct evidence that an atmosphere of gradually increasing thickness partially obscured the light from the star. Ganymede's atmosphere was thought to be very thin, based on the amount of light gradually extinguished during the occultation. Its proposed constituents were methane (CH_4) and ammonia (NH_3). Surface pressure on Ganymede was estimated as at least 10^{-3} mb but not >1.0 mb.

Voyager 1 studies failed to record any evidence of an atmosphere on Ganymede. The UV spectrometer was capable of detecting an extinction of as little as 10 percent. This places an upper limit of the surface density of any existing atmosphere at 6×10^{-8} cm^{-3}. Revised estimates place the surface atmospheric pressure at an upper limit of only 10^{-8} mb.

Magnetosphere

A series of magnetospheric disturbances in the region between the Jovian magnetosphere and Ganymede were observed by Voyager 2. The nature and source of these perturbations may be Ganymede, especially since they have been observed near and equidistant from Jupiter relative to Ganymede. The data are inadequate to explain these phenomena at the present time.

Atmospheric Temperature

Surface temperatures on Ganymede are $\simeq 130^{\circ}K$, lying between the minimum ($80^{\circ}K$) and the maximum ($155^{\circ}K$) observed for the Galilean satellites.

Summary

There are many unanswered questions concerning Ganymede. The Voyager spacecraft made significant discoveries concerning the surface geology of Ganymede, however, atmospheric data are lacking, as is evident from the brief treatment given Ganymede in this chapter.

References

Butterworth, P.S. et al., An upper limit to the global SO_2 abundance on Io, Nature, 285, 308 ff., 1980.

Campbell, D.B. et al., Galilean satellites of Jupiter: 12.6 centimeter radar observations, Science, 196, 650 ff., 1977.

Carlson, R.W. et al., Atmosphere on Ganymede from its occultation of SAO 186800 on 7 June, 1972, Science, 182, 53 ff., 1973.

Carr, M.H. et al., Volcanic features of Io, Nature, 280, 729 ff., 1979.

Consolmagno, G.J., Sulfur volcanoes on Io, Science, 205, 397 ff., 1979.

Cook, A.F. et al., Dynamics of volcanic plumes on Io, Nature, 280, 743 ff., 1979.

Fanale, F.P. et al., Io: a surface evaporite deposit, Science, 186, 922 ff., 1974.

Fanale, F.P. et al., Significance of absorption features in Io's IR reflectance spectrum, Nature, 280, 763 ff., 1979.

Garnett, D.A. et al., Auroral hiss observed near the Io plasma torus, Nature, 280, 767 ff., 1979.

Gold, T., Electrical origin of the outbursts on Io, Science, 206, 1071 ff., 1979.

Goldstein, R.M. and Morris, G.A., Ganymede: observations by radar, Science, 188, 1211 ff., 1975.

Goldstein, R.M. and Green, R.R., Ganymede: radar surface characteristics, Science, 207, 179 ff., 1980.

Johnson, T.V. et al., Volcanic resurfacing rates and implications for volatiles on Io, Nature, 280, 746 ff., 1979.

Kumar, S., The stability of an SO_2 atmosphere on Io, Nature, 280, 758 ff., 1979.

Matson, D.L. et al., Images of Io's sodium cloud, Science, 199, 531 ff., 1978.

Morrison, D. et al., Photometric evidence on long-term stability of albedo and colour markings on Io, Nature, 280, 753 ff., 1979.

Nash, D.B. and Nelson, R.M., Spectral evidence for sublimates and adsorbates on Io, Nature, 280, 763 ff., 1979.

Nelson, R.M. et al., Io: longitudinal distribution of sulfur dioxide frost, Science, 210, 784 ff., 1980.

Peale, S.J. et al., Melting of Io by tidal dissipation, Science, 203, 892 ff., 1979.

Pearl, J. et al., Identifications and gaseous SO_2 and new upper limits for other gases on Io, Nature, 280, 755 ff., 1979.

Sagan, C., Sulfur flows on Io, Nature, 280, 750 ff., 1979.

Smith, B.A. et al., The role of SO_2 in volcanism on Io, Nature, 280, 738 ff., 1979.

Smythe, W.D. et al., Spectral evidence for SO_2 frost or adsorbate on Io's surface, Nature, 280, 766 ff., 1979.

Strom, R.G. et al., Volcanic eruption plumes on Io, Nature, 280, 733 ff., 1979.

Whitteborn, F.C. et al., Io, an intense brightening near 5 micrometers, Science, 203, 643 ff., 1979.

CHAPTER VII

SATURN AND TITAN

SATURN

Descriptive Statistics:

Distance from Sun	9.5 A.U.
Orbital Tilt	29.5 years
Revolution Period	10.2 years
Axial Tilt	$26^{\circ}45'$
Mass	95 times that of earth
Equatorial Radius (R_S)	60,000 km
Polar Radius	≃10% less than the equatorial R_S
Density	0.7 g cm^{-3}
Ring System Radius	≃500,000 km

Introduction

Saturn is the most magnificent object in the solar system when viewed telescopically. Its yellowish hues and extensive ring system make it a favorite object of amateur astronomers. Yet this easily observed planet is so distant from the earth (1.125×10^9 km) that scientific data were extremely limited until the Pioneer and Voyager probes. A recent photograph of Saturn by Voyager 1 is presented in Fig. 7.1.

There are many similarities between Jupiter and Saturn (Table 7.1). Their planetary radii are quite large in comparison to the earth's radius. They both have extensive satellite systems. Jupiter and Saturn have 16 and 14 moons, respectively with the distinct possibility additional moons may be discovered in Saturn's rings. Both Saturn and Jupiter rotate rapidly on their axes and have planetary masses far in excess of the earth's mass. The mean densities for Saturn and Jupiter are 1.3 and 0.7 g cm^{-3}, respectively, much less than the value for the earth. All three planets have magnetic fields but those of Saturn and Jupiter are significantly greater in intensity and spatial distribution. While the earth, Jupiter, and Saturn experience polar flattening due to

Fig. 7.1. Voyager 1 photograph of Saturn from 106,250,000 km taken on August 24, 1980. The dark band along the equator is the shadow from the rings. Three of Saturn's moons are visible in this photo.

TABLE 7.1

A comparison of the characteristics of Saturn, Jupiter, and the Earth

	Saturn	Jupiter	Earth
Distance from the Sun (A.U.)	9.5	5.2	1.0
Radius (km)	60,000	71,400	6,380
Length of Year	29.5	11.9	1.0
Length of Day	9 hr, 55 min	10 hr, 40 min	23 hr, 56 min
Mass	95	318	1.0
Density (g cm^{-3})	0.7	5.2	5.5
Mean Distance to Outer Edge of Magnetosphere (in Radii)	30 R_S	99 R_J	10 R_E
Number of Satellites	14	16	1

their rotation, Saturn is unusual because the polar flattening results in a 10 percent difference between the planet's polar and equatorial radius. Saturn's mid-latitude regions are depressed inward (concave) by over 120 km, a far greater effect than observed on any other planet. In general, there are a number of striking similarities between Saturn and Jupiter as well as some unique differences.

Atmospheric Composition

The composition of Saturn's atmosphere is much like Jupiter's atmosphere, Hydrogen, deuterium (heavy hydrogen), ammonia, methane, ethane, ethylene, acetylene, and phosphine have all been identified in Saturn's atmosphere. Helium is probably present deeper within the atmosphere. Voyager photography in the ultraviolet wavelengths revealed a well defined banded pattern across the planet's disk but the turbulence so strikingly visible on Jupiter and present on Saturn is not as evident in the photographs. Prior to the Voyager missions, scientists had suggested that Saturn had turbulent vortices as part of its atmospheric structure but that they were hidden beneath a thick layer of atmospheric haze. The rationale for the thick haze layer was based on Saturn's temperature of $140^{o}K$ at the 1 bar reference level. Saturn is $\simeq 30^{o}K$ colder than Jupiter at the 1 bar level. Colder temperatures suggest that Saturn's cloud deck is found at lower levels relative to the reference level.

The base of the clouds would also be lower since the level at which the vapor pressure of a gaseous constituent would equal its partial gas pressure in the atmosphere (i.e. saturation levels) would be at greater depths (and pressures) on Saturn. For example, ammonia clouds which exist down to the 600 mb level on Jupiter extend downward to the 1-3 bar level on Saturn. The composition of the thick haze was presumed to be ammonia crystals which had been vertically convected aloft to pressure levels comparable to those on Jupiter. Since these ammonia crystals originate from a cloud deck at lower altitudes, the haze layer appeared to be significantly thicker on Saturn. The thick haze layer explains why earth based photography and the earlier Pioneer probe were only able to record the faint outlines of the cloud zones and belts which we now know exist on Saturn. We shall discuss Saturn's clouds after first considering the atmospheric temperature structure and energy budget of the planet.

Temperature

The vertical temperature profile from the 1 bar (1,000 mb) reference level to the 0.1 mb level is presented in Fig. 7.2. The temperature at the reference level is $\approx 140^{o}$K and it decreases vertically to $\approx 120^{o}$K at the cloud tops, which are at or near the 650 mb level. The tropopause is clearly visible at the 100 mb level where the temperature is $\approx 85^{o}$K. The region above 100 mb coincides with the upper regions of the haze layer. The region below 100 mb coincides with the more convective region of the atmosphere. Temperature measurements derived from modelling studies suggest that a temperature difference of 2.5^{o}K exists between the belts and zones near the equator and that this temperature difference continues to the tropopause (100 mb).

Saturn may also have a thin ionosphere layer. Radio occultation data revealed the existence of two regions at altitudes of 2,200 km and 2,800 km where observations reported increased electron concentrations. The greatest concentration of electrons was at the 2,800 km altitude and it seems reasonable to assume that an ionosphere layer exists at this altitude.

A limited latitudinal temperature profile is presented in Fig. 7.3. The upper curve, sensing at 20 μm wavelengths depicts the latitudinal temperature profile from the 100 mb to 60 mb level. The

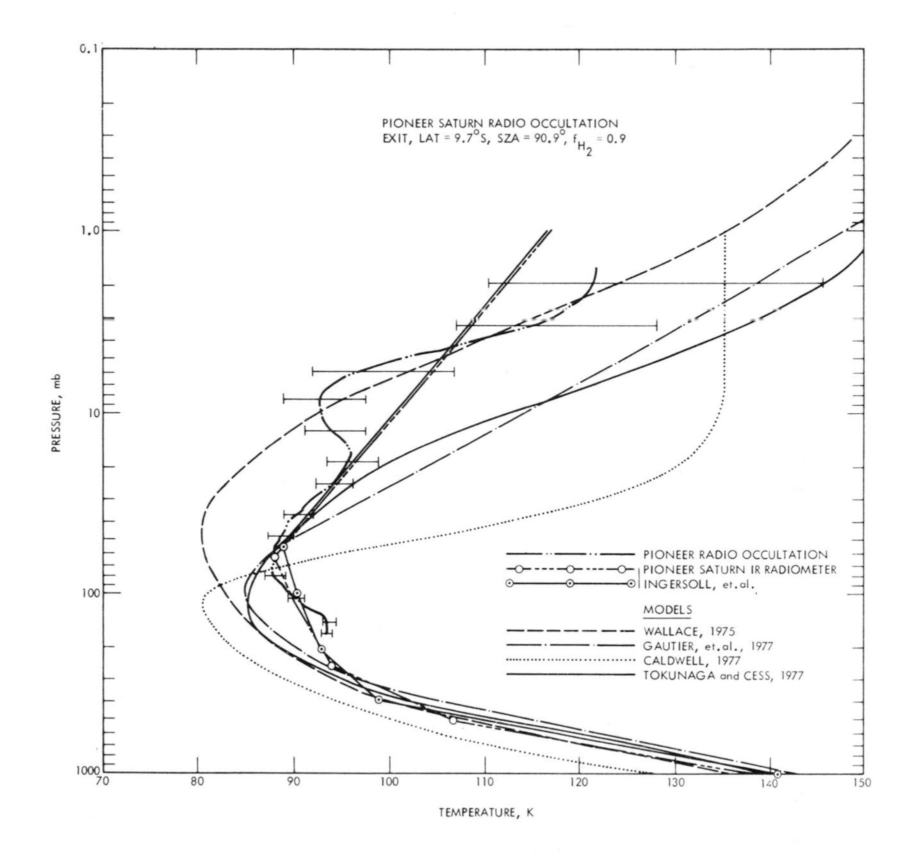

Fig.7.2. An atmospheric temperature/pressure profile of Saturn's atmosphere down to the 1 bar level. Courtesy of A.J. Kliore et al., J. Geophys. Res.,85,p.5868,Nov.,1980. Copyrighted by the American Geophysical Union.)

data derived from the 45 μm sensor depict the temperatures in the 600 mb to 100 mb region. Both profiles show a colder equatorial zone (7°N to 7°S). The regions from 7°N to 12°N and 7°S to 13°S are warm belts. Similar patterns of temperature increases and decreases prove, along with the Voyager imagery, that the belt/zone pattern extends poleward to the mid-latitude region. The large scale belt/zone features break down into a more mottled or disturbed region of waves and eddies from the mid-latitudes to the poles.

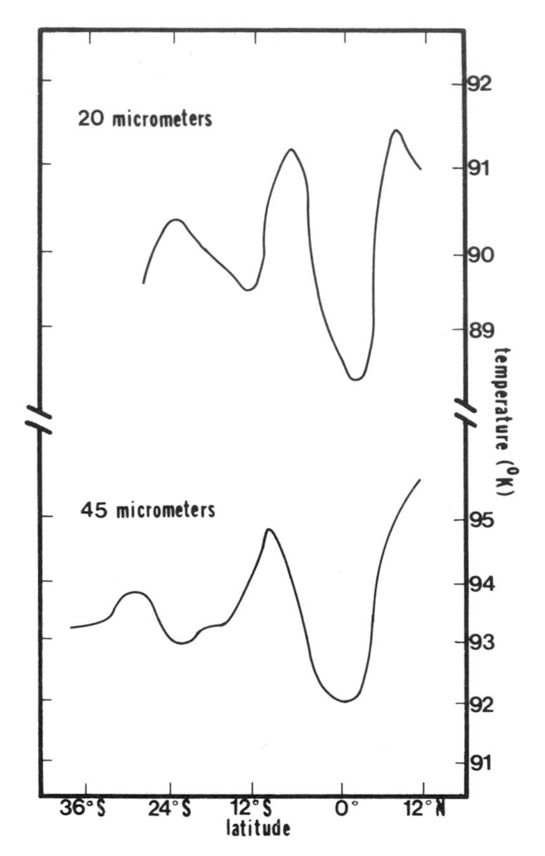

Fig. 7.3. Latitudinal temperature profiles for Saturn at 20 μm and 45 μm. Temperatures correspond to the 500-600 mb level. Note the cold equatorial zone (7°N-7°S) and the warmer belts to the north and south. (Courtesy A.P. Ingersoll et al., Science, v 207, p.441, 25 Jan., 1980. Copyright 1980 by the American Association for the Advancement of Science.)

The Energy Budget

Saturn receives ≃3.5 times less solar energy than Jupiter. Saturn radiates ≃2.2 times more energy to space than it receives from the sun assuming an albedo of 0.45. Unlike Jupiter however, Saturn's energy surplus cannot be solely attributed to or explained by gravitational contraction processes and the residual heat from the planet's formation. The explanation for the observed heat flux from Saturn will be more easily understood if we postpone this topic until we have examined the planet's magnetic characteristics.

Fig. 7.1. Voyager 1 photograph of Saturn from 106,250,000 km taken on August 24, 1980. The dark band along the equator is the shadow from the rings. Three of Saturn's moons are visible in this photo.

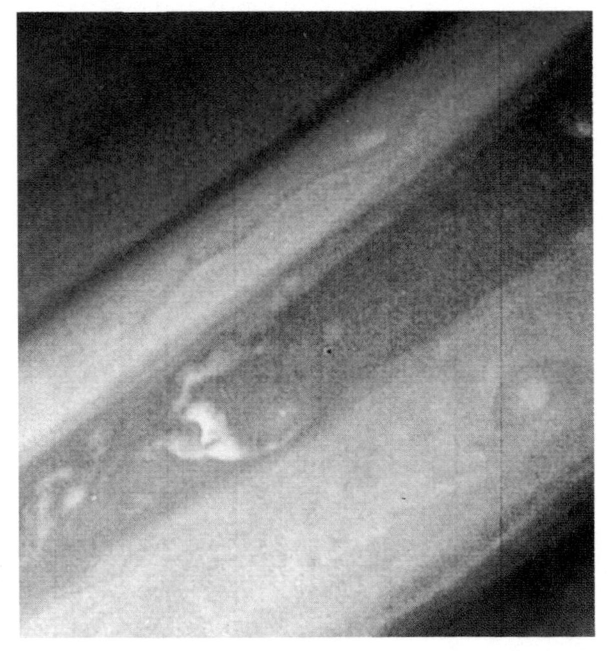

Fig. 7.7. An enhanced color image of Saturn's northern hemisphere taken by Voyager 1 on November 5, 1980 at a range of 9 million km. Small scale convective clouds are visible in the brown belt. The smallest features visible are 175 km across. (Courtesy Jet Propulsion Laboratory/NASA.)

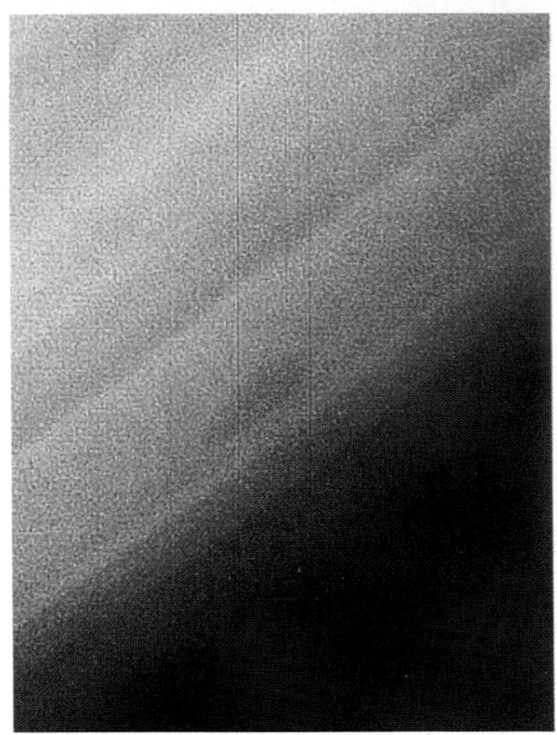

Fig. 7.8. Saturn's Great Red Spot located at 55°S. Photograph by Voyager 1 taken on November 6, 1980 at a range of 8.5 million km. (Courtesy Jet Propulsion Laboratory/NASA.)

Fig. 7.9. Two brown ovals 6,000 km across located at 40°N and 60°N. The photograph was taken by Voyager 1 from a range of 7.5 million km. The polar oval's structure is similar to Saturn's red spot at 55°S. (Courtesy Jet Propulsion Laboratory/NASA.)

The Magnetosphere

Scientists had thought that Saturn possessed an intrinsic planetary magnetic field for some time, however, the suspense heightened as the Pioneer 10 spacecraft approached closer and closer to Saturn without any evidence of a bow shock. Finally, the first bow shock crossing occurred at 24 Saturn radii (R_S), the second at 23 R_S, and the third at 20 R_S. The magnetopause was crossed at 17 R_S. The strength of Saturn's magnetic field, especially in comparison to Jupiter's, was significantly smaller than had been predicted for Saturn. Saturn's smaller magnetic field is still many times greater than the earth's magnetic field and is actually intermediate between that of Jupiter and the earth. The polarity of their (Saturn and Jupiter) magnetic fields is opposite to the polarity of the earth's magnetic field. All three planets have dipolar fields but only Saturn's magnetic field is aligned to within 1° of its rotational axis as it corotates rapidly with the planet.

Saturn's magnetosphere responds to changes in solar wind intensity quite readily. The solar wind, in turn, is slowed to one third of its original velocity while plasma temperatures increase from 30,000 to 5×10^{5} ^oK in the magnetosheath. Saturn's trapped radiation belts are comparable in intensity to the earth's Van Allen belts although they occupy a region ten times larger.

The magnetosphere may actually be subdivided into four regions: the outer magnetosphere, the slot, the inner magnetosphere, and the rings. The outer magnetosphere extends from 17 to 7.5 R_S and contains the corotating plasma composed of ionized atomic oxygen and hydroxide. The slot represents an area swept free of particles by three of Saturn's moons: Dione, Tethys, and Enceladus. The slot extends from 7.5 to 4.0 R_S. The inner magnetosphere, inside 4.0 R_S, contains extremely high energy protons and electrons. These particles are nearly absent from the rings, no doubt due to their absorption by the ring materials. The four subdivision of the magnetosphere comprises the ring region out to at least 30 R_S, however, before proceeding to a discussion of the rings and ring materials, we should, at this point, return to the topic of Saturn's internal source of heat energy and its relationship to the magnetosphere.

Saturn's magnetosphere is a product of its intrinsic magnetic field. The intrinsic magnetic field reveals the existence of and, to an extent, defines the dimensions of, an electrically conducting, liquid core region of metallic hydrogen. The surplus heat and magnetic field point to a unique source for the surplus heat energy that is radiated out to space.

Saturn reradiates all the solar energy it absorbs. Estimates suggest that one third of the heat reradiated from Saturn is from the residual heat associated with the planet's formation. Saturn radiates too much infrared heat energy and therefore requires an additional process to explain the source of this surplus energy. The proposed process is referred to as differentiation. Differentiation not only explains the source of extra heat but also limits the liquid metallic hydrogen region to deeper within the planet. The smaller liquid hydrogen core extends from 0.2 to 0.5 R_S while the inner core extends from the center to 0.2 R_S. The inner core probably consists of various oxides of magnesium (MgO), iron (FeO), and silica (SiO_2) along with iron sulfide (FeS). The smaller dimensions of the liquid metallic hydrogen region clearly explain the smaller than expected magnetic field.

The surplus energy radiated by Saturn is derived from the gravitational energy released as denser helium moves toward the interior of the planet. Helium raindrops grow as the helium becomes insoluble at temperatures $>10^4$ °K and pressures $\approx$2,000 bars. The droplets grow to 1 cm in diameter and fall due to gravity. They dissolve at greater depths where temperatures are higher. This results in an outer layer of the planet, including the atmosphere, that is depleted of helium and a new differentiation model for the internal structure of Saturn (Fig. 7.4). Thus helium differentiation not only explains Saturn's extra heat energy source but also the reason for the weaker magnetic field (i.e. a smaller metallic core region).

It is probable that the dimensions of Saturn's magnetosphere may expand significantly during those periods when Jupiter is aligned between the sun and Saturn. The Jovian magnetosphere would deflect the solar wind at its bow shock. Saturn, orbiting through Jupiter's vast magnetotail, should experience a reduction or total cutoff of the solar wind. The data from such an event would provide us with

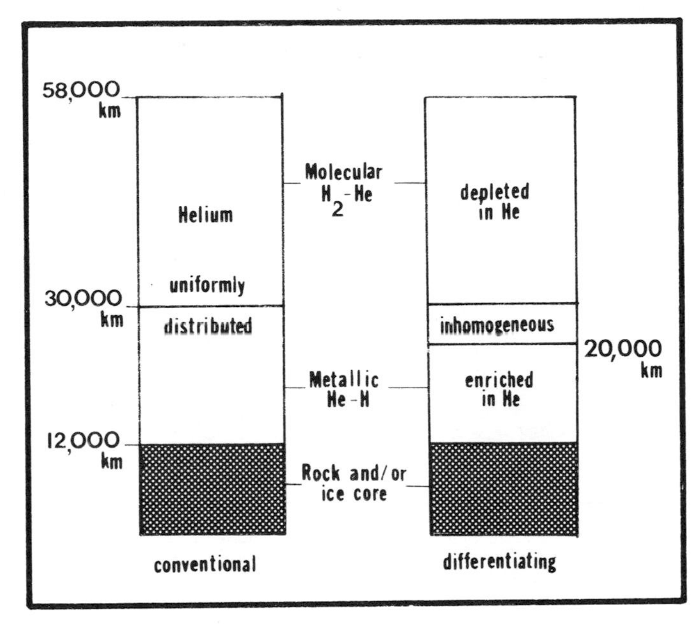

Fig. 7.4. Conventional and differentiating models of Saturn's interior. The differentiating model has an inhomogeneous layer where helium raindrops form and release energy. (Adapted from and courtesy of D.J. Stevenson, Science, v 208, p.747, 16 May 1980. Copyright 1980 by the American Association for the Advancement of Science.)

a rare opportunity to study the behavior of an expanding and contracting magnetosphere. We shall soon have the opportunity to view such data as the Voyager 2 spacecraft is expected to encounter Saturn in August, 1981 at a time when Saturn will be protected by Jupiter's magnetotail.

Saturn's Rings

Saturn's ring system lies within Saturn's magnetosphere, otherwise a discussion of Saturn's ring system would be beyond the scope of this book. Normally, we have observed the rings in reflected sunlight from a line of sight position in front of the rings, i.e. the earth. The orbital trajectory of the Pioneer probe provided the first images taken from a position above and to one side of the rings so that the rings were backlighted by the sun. A similar orbital trajectory resulted in the magnificent Voyager photographs and innumerable new discoveries concerning the structure of the ring system. Prior to the Voyager 1 discoveries, the ring system was considered as a series of orbiting bands of particles separated by "empty" regions called divisions (Table 7.2).

TABLE 7.2

Distances of rings and division from Saturn prior to Voyager 1

		Distance from "Surface" of Saturn (km)	R_S
D Ring			
Inner		near cloud tops	
Outer		8,600	1.14
	Guerin Division		
C Ring			
Inner		12,600	1.21
Outer		30,000	1.50
	French Division		
B Ring			
Inner		31,800	1.53
Outer		57,000	1.95
	Cassini Division		
A Ring			
Inner		61,800-73,050	2.00-2.22
	Encke Division		
Outer		73,950-77,400	2.23-2.29
	Pioneer Division		
F Ring			
Inner		80,160-82,260	2.34-2.37
G Ring		168,000	2.80

The Pioneer spacecraft added to our knowledge of Saturn's ring system. It revealed the existence of the F ring located 3,600 km from the A ring. The spacecraft measured a thousandfold decrease in the intensity of the background radiation in a distance of 0.01 R_S as it passed by the outer edge of the A ring on its way into the ring system. This absorption of high energy particles by the ring materials makes the ring system one of the most radiation free regions in the solar system. The Pioneer probe discovered that the Cassini Division appeared bright when viewed from a backlighted position. This proved that the division was not empty but contained at least enough particles to scatter sunlight effectively. The B ring appeared bright but with a patchy pattern. This was the result of light leaks through those portions of the B ring where particle densities were much lower. Pioneer observed the French Division (1,800 km) which separated the C ring from the B ring. It observed that the C ring consisted of very few, low density ice particles. The C ring's particle concentrations were estimated to be greater than, but similar to, the number of particles in the Cassini Division. The D ring was observed as nearly "touching" the upper atmosphere at 1.14 R_S and it was considered possible that the D ring might be an extension of the ammonia haze particles from the upper atmospheric haze layer. The G ring, the last of Pioneer's ring discoveries, was discovered at a distance of 2.8 R_S.

A faint hydrogen glow observed in conjunction with the rings suggested that the hydrogen was either from Titan's atmosphere or was produced as the water-ice ring materials dissociated after bombardment and absorption of high energy protons from the magnetospheric radiation belts. The diameters of ring particles were computed as ranging from 1.5-15 cm. Ring temperatures were 70°K on the sun side and 55°K on the opposite side. These Pioneer discoveries were impressive additions to the body of scientific knowledge of our solar system but they have been completely overshadowed by the Voyager imagery and discoveries.

Voyager 1 discovered that the rings of Saturn were, in fact, 500-1,000 smaller rings. These rings are distributed across the ring plane and make the more traditional A-F ring classification (Table 7.2) obsolete. It is impossible to map all the rings but a revised diagram presents a more accurate picture of the ring system around Saturn (Fig. 7.5). The composition of the ring particles appears to be primarily water-ice, some dust, and an occasional larger diameter particle spaced intermittently throughout. The discovery of radial spikes within the ring structure verified two earlier earth based observations of these phenomena. The exact process producing these radial spikes is unknown at this time but is thought to be due, in part, to magnetospheric interactions with

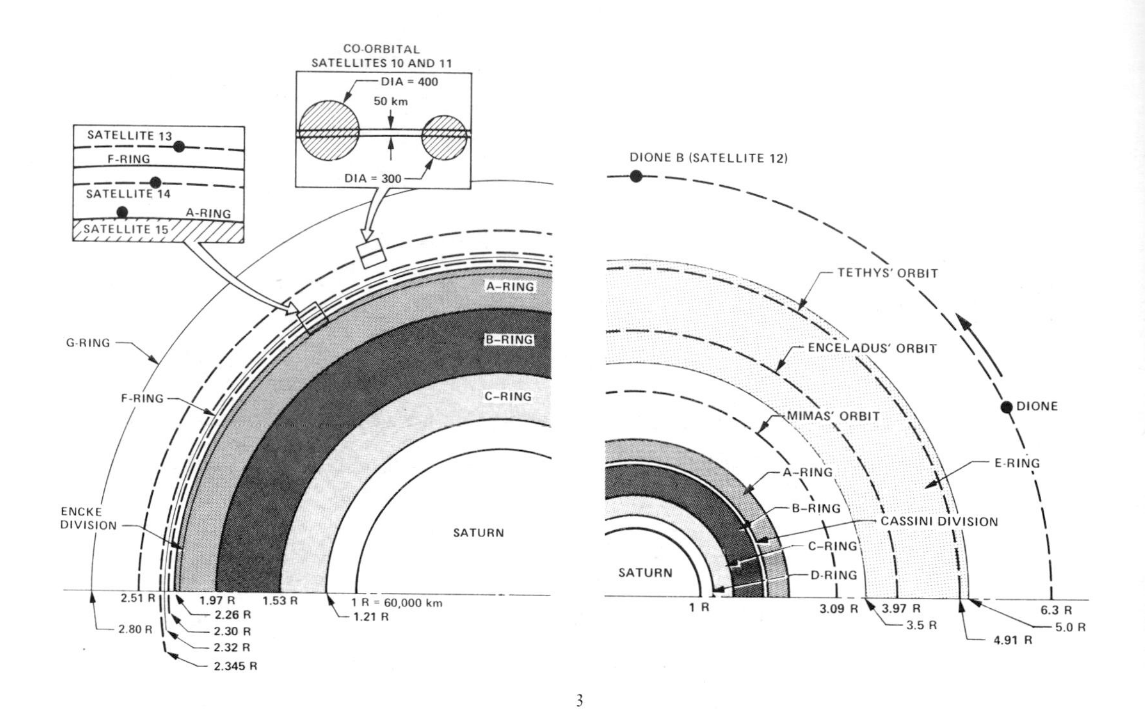

Fig. 7.5. A revised view of the major components of Saturn's ring system after Voyager 1. (Courtesy Jet Propulsion Laboratory/NASA.)

some of the ring particles or gravitational forces. The radial spokes are probably not groups of particles but regions where particle concentrations are very low. Fewer particles would reflect less sunlight than other portions of the ring system and appear as the darker spokes. The absence of particles might be explained by the gravitational influences from several of Saturn's satellites. If the spoke presence is not gravitationally induced, then a search for some unknown magnetospheric interaction should provide an explanation for their presence within the ring structures.

One of the interesting aspects of the Voyager results was the discovery that the F ring is actually two or three rings in a braided pattern (Fig. 7.6). The cause for this braided pattern is

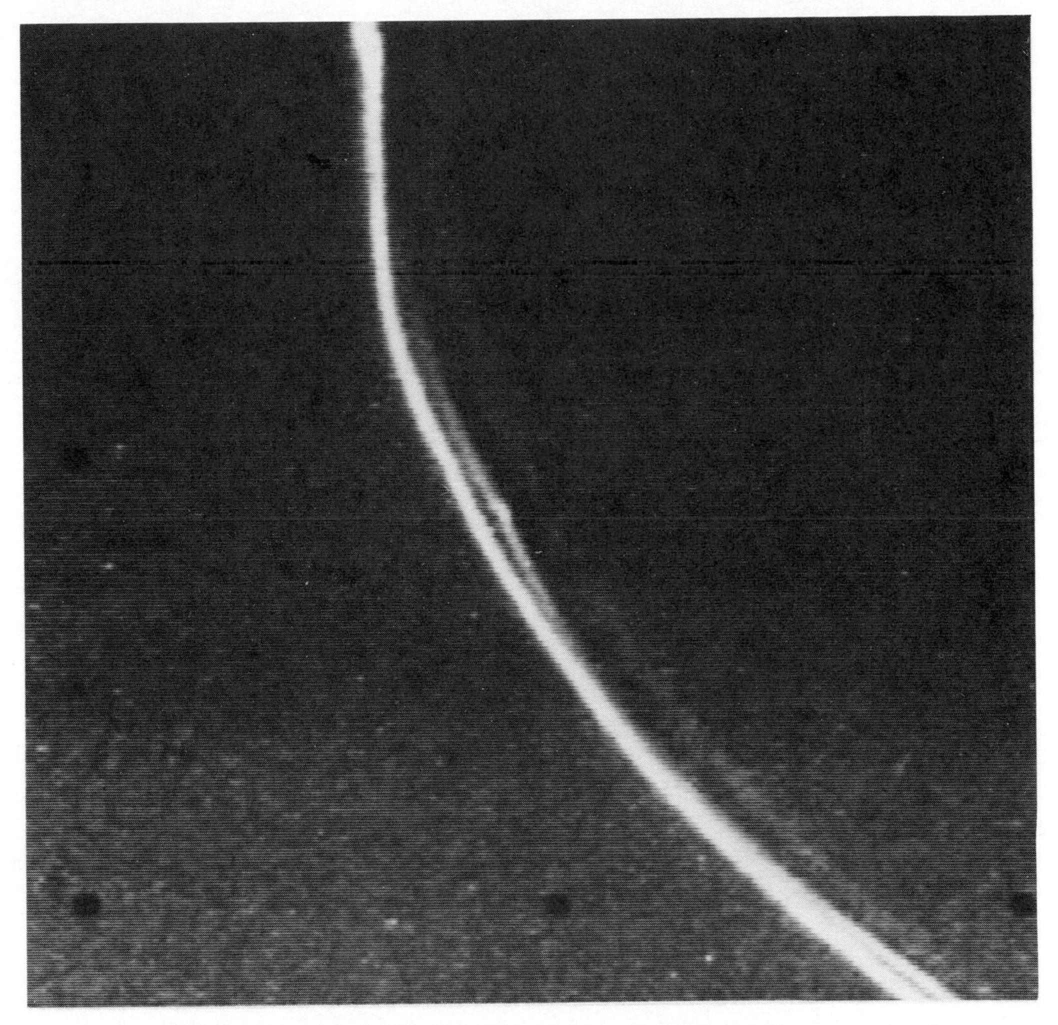

Fig. 7.6. Saturn's braided F ring photographed by Voyager 1 at a range of 750,000 km. Two bright and one faint strand show a complex braided pattern. The bright "knot" (top) are either local clumps of ring material or "mini-moons." (Courtesy Jet Propulsion Laboratory/NASA.)

unknown. The F ring also appears knotted in places (top - Fig. 7.6). These knots may be clumps of ring materials or possibly "mini-moons." Several small diameter moons were accidently discovered within the ring system because the Voyager imagery was subjected to such intense scrutiny by the NASA scientists.

Saturn's Clouds

The most significant contributions made by Voyager were the detailed images of Saturn's cloud layers. The results of these images have demonstrated that, beneath the haze, Saturn's cloud patterns of belts and zones is very similar to the patterns found on Jupiter. The terminology applied to naming Saturn's cloud bands is the same as that used in naming Jupiter's bands, i.e. EZ-equatorial zone, NEB-north equatorial belt, NTrZ-north tropical zone, NTeZ-north temperature zone, mid-latitudes, and polar regions. The patterns repeat in the southern hemisphere. The cloud bands are symmetrical from the equator to the polar regions (Fig. 7.7). Convective phenomena appear as the lighter areas in this enhanced color image of Saturn's northern hemisphere. An unusual convective cloud surrounded by a dark collar is also visible (middle right - Fig. 7.7).

Saturn's southern hemisphere mirrors the northern hemisphere except for the discovery of a Great Red Spot (GRS) 12,000 km in length at 55°S (Fig. 7.8). Saturn's red spot is smaller than Jupiter's red spot but it is quite probable that the processes which produce the GRS on Jupiter are also responsible for the GRS on Saturn. The large size and southern hemispheric location coupled with similar atmospheric gaseous constituents (phosphine) support the idea that both red spots result from similar causes. The only factor that might suggest the need for an alternate cause is their latitudinal position. Jupiter's GRS is at 23°S while Saturn's GRS is at 55°S. Perhaps the release of energy due to the helium differentiation is more than adequate to support a storm of the red spot type at higher latitudes on Saturn. Only future research will confirm the similarities between the two red spots.

Two brown ovals were discovered in Saturn's northern hemisphere (Fig. 7.9). Both are 10,000 km in length and are located at 40°N and 60°N. The brown oval at 60°N has a structure similar to Saturn's GRS at 55°S. The color difference may result from changes in the atmospheric chemistry which control the coloration of the cloud belts, zones, and spots.

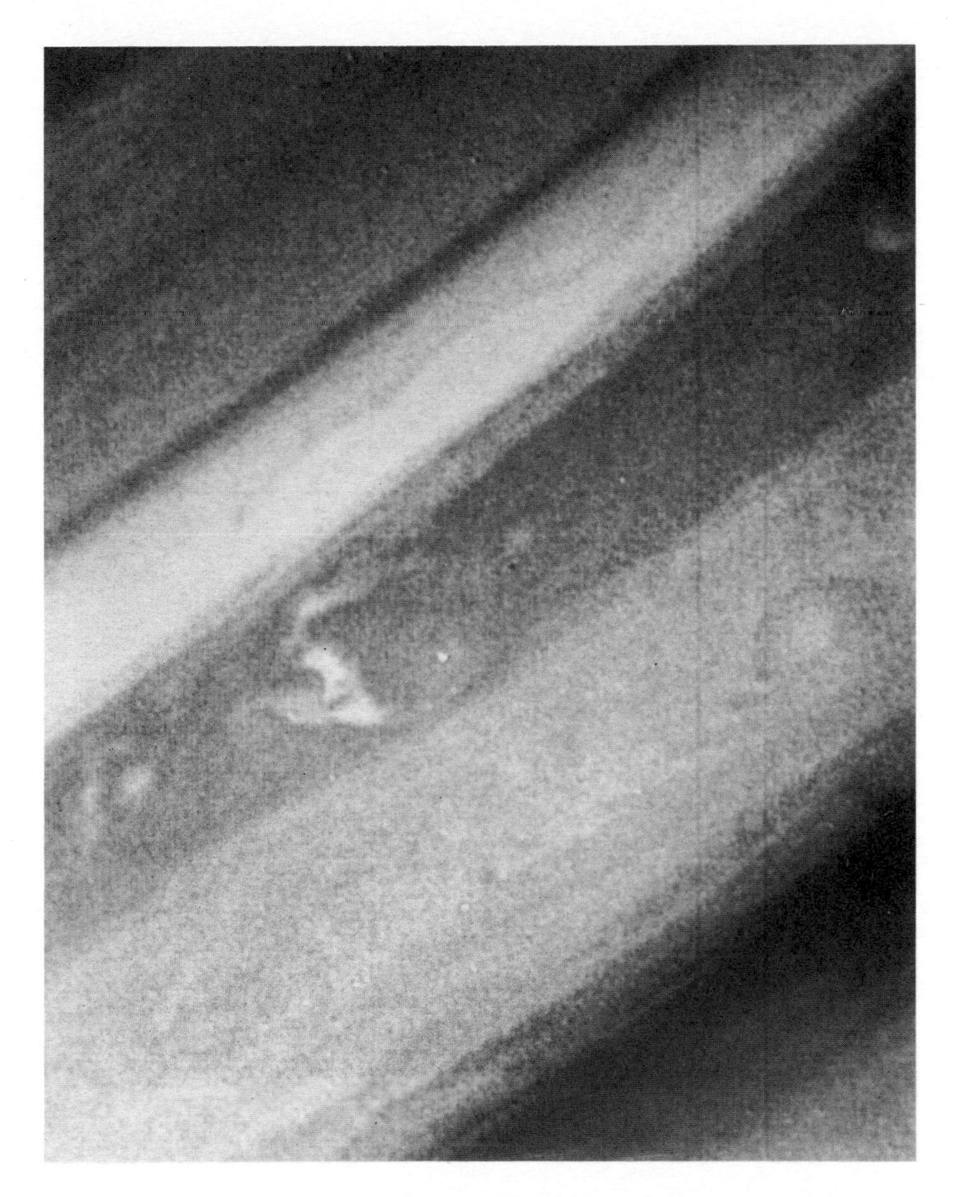

Fig. 7.7. An enhanced color image of Saturn's northern hemisphere taken by Voyager 1 on November 5, 1980 at a range of 9 million km. Small scale convective clouds are visible in the brown belt. The smallest features visible are 175 km across. (Courtesy Jet Propulsion Laboratory/NASA.)

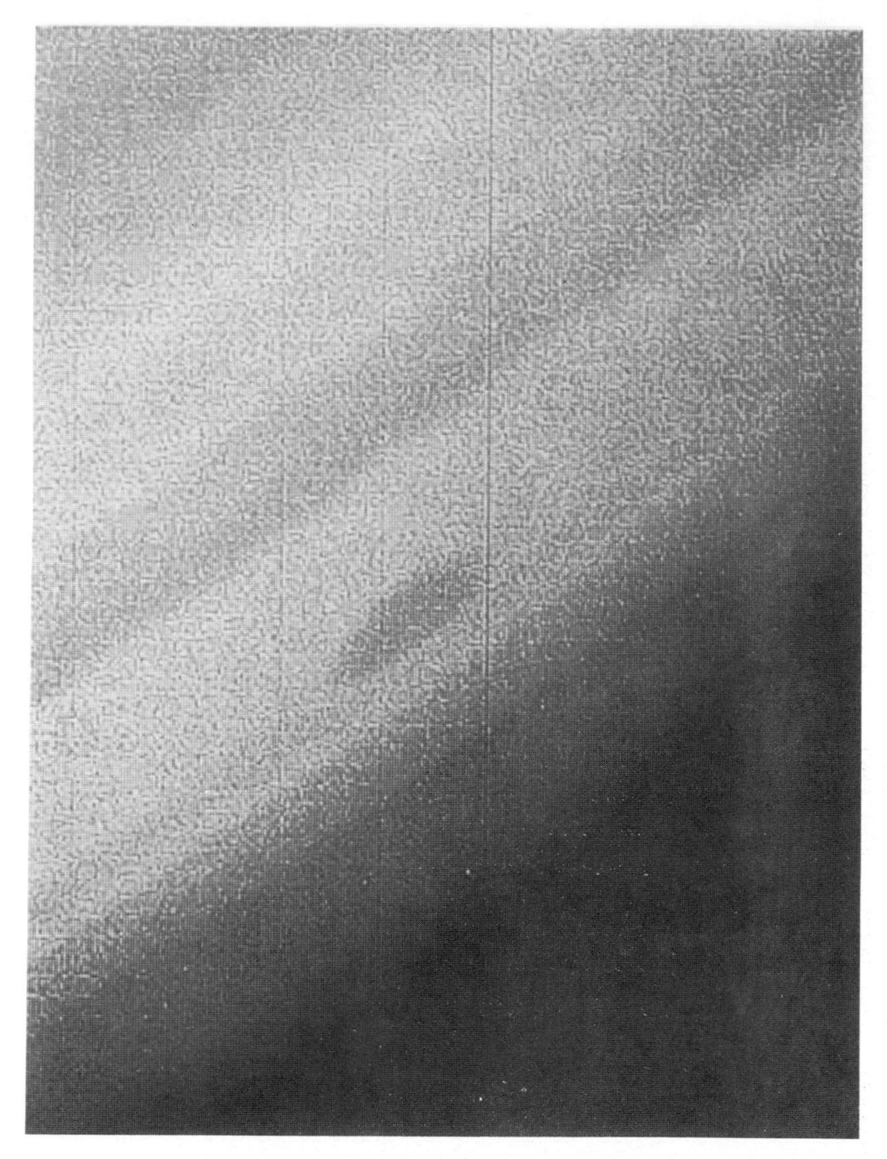

Fig. 7.8. Saturn's Great Red Spot located at 55°S. Photograph by Voyager 1 taken on November 6, 1980 at a range of 8.5 million km. (Courtesy Jet Propulsion Laboratory/NASA.)

Saturn's polar clouds are very similar to Jupiter's polar clouds (Fig. 7.10). The belts and zones give way to waves and turbulent eddies poleward of 60° due to the Coriolis effect. There is little

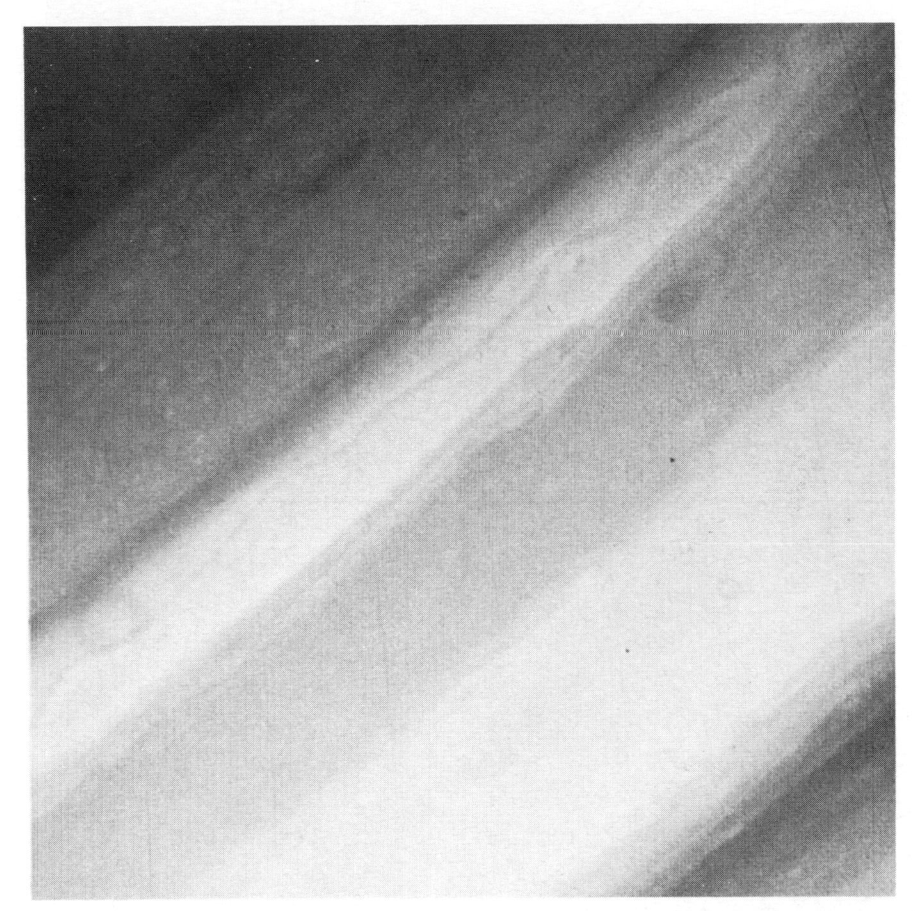

Fig. 7.9. Two brown ovals 6,000 km across located at 40°N and 60°N. The photograph was taken by Voyager 1 from a range of 7.5 million km. The polar oval's structure is similar to Saturn's red spot at 55°S. (Courtesy Jet Propulsion Laboratory/NASA.)

order to these small scale features. Similar phenomena exist in the north polar regions where chevron-type patterns of white features delineate the local cloud motion (Fig. 7.11). Wind speeds are measured by the rates at which these small scale features change. The smallest features visible are only 60 km.

Winds

Jet streams are also present on Saturn. Several have been identified at the boundaries between the equatorial region and the

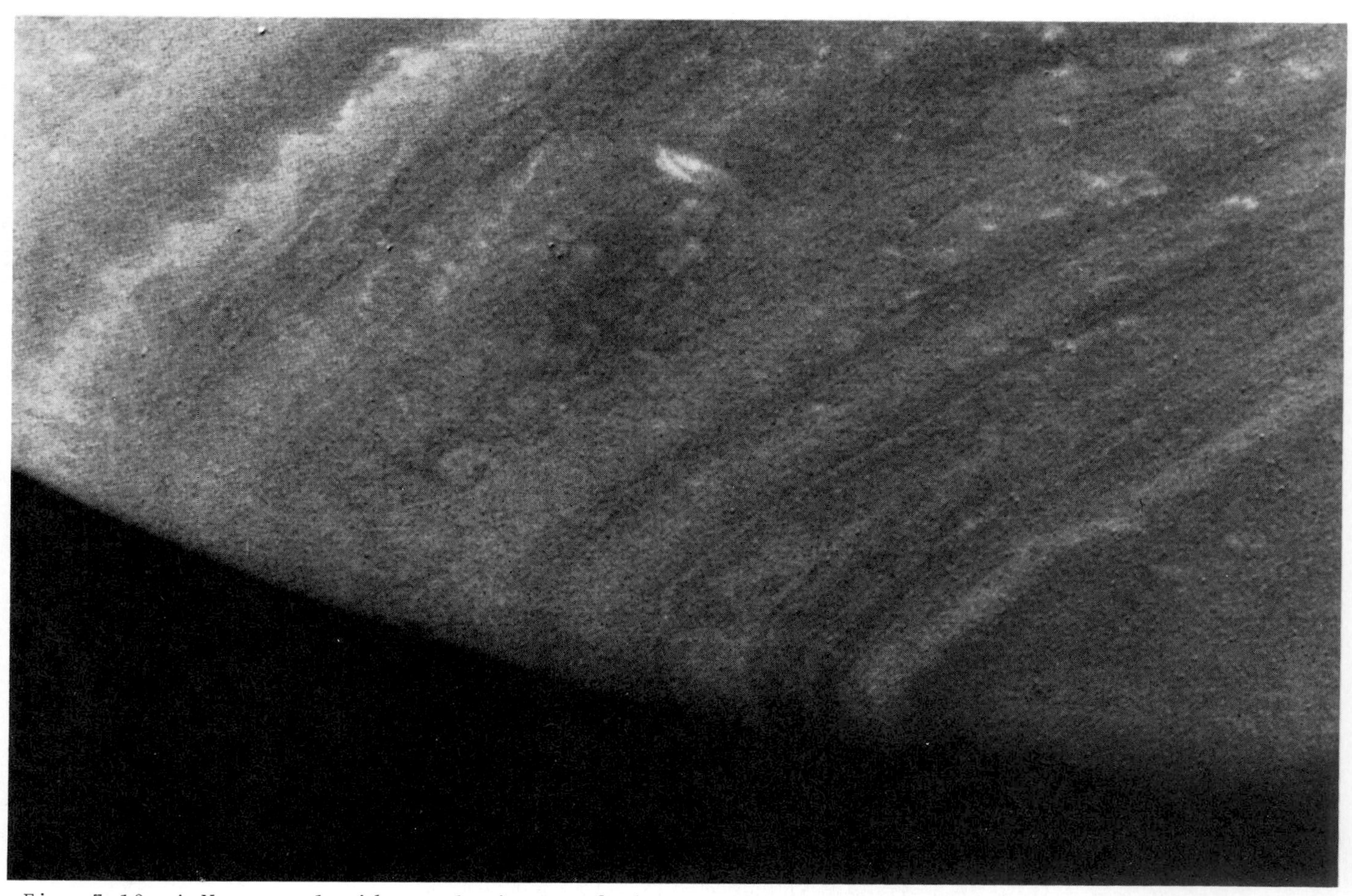

Fig. 7.10. A Voyager 1 wide-angle image of the numerous small cloud features of Saturn's south polar region and mid-latitudes. This image was taken on November 12, 1980 from a distance of 442,000 km. (Courtesy Jet Propulsion Laboratory/NASA.)

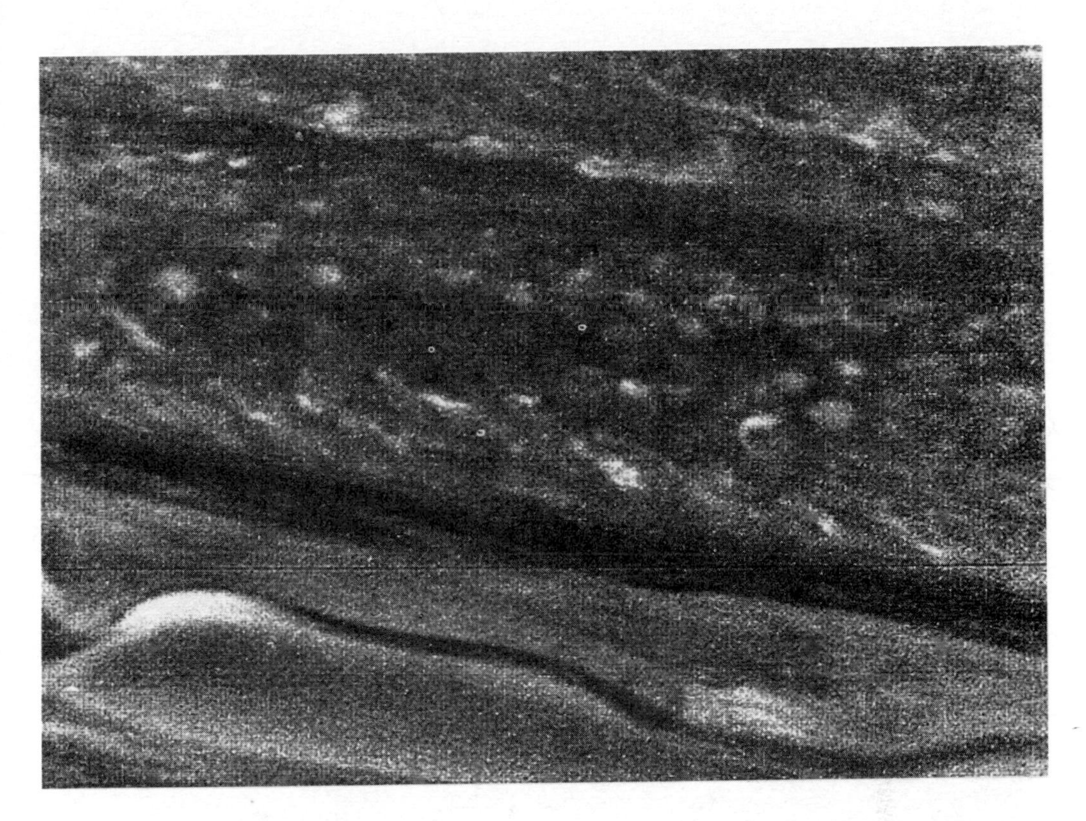

Fig. 7.11. A computer enhanced Voyager 1 image of the small scale cloud features in Saturn's north polar region. The Chevron-type pattern of white features indicates local cloud motion. This photograph was taken on November 10, 1980 from 3 million km. The smallest features are 60 km across. (Courtesy Jet Propulsion Laboratory/NASA.)

north tropical zone and at 70°N. The evidence for these jet streams is based on the wavy appearance and lack of contrast at the belt/zone boundaries and their resemblance to known jet stream boundaries on Jupiter. The velocities of Saturn's jet streams have been estimated as ≃1,400 km hr^{-1} making them the fastest jet streams measured anywhere in our solar system.

Aurora and Radio Emissions

There is some evidence that particle precipitation down into

Saturn's atmosphere produces auroral displays, however, these auroral events have not been verified as yet. Similarly, weak radio emissions have been reported from Saturn but the source or the process responsible for radio wave generation is, as yet, unidentified. The bursts are cyclical in nature and are definitely non-thermal radio noise. The frequency of these bursts is close to 200 kHz. It is virtually impossible to monitor these emissions from the earth because these emissions at ≈200 kHz are lost and/or obscured by a radio band used for communication on earth.

Summary

Saturn is a complex planet. All the data suggest that, although it is physically smaller than Jupiter, many similarities exist between the two gaseous giants. The initial Voyager discoveries have shown Saturn as a dynamic, entity with its own set of unique characteristics. Certainly Saturn (and its ring system) remains the most beautiful object in our solar system.

TITAN

Descriptive Statistics:

Radius	2,400 km (as large as Mercury)
Distance from Saturn	500,000 km
Distance from the Sun	9.5 A.U. (same as Saturn)
Mass	1.9 times the earth's moon

Atmospheric Composition

Titan is the largest satellite of Saturn. It is 4,800 km in diameter and orbits Saturn at a distance of 500,000 km in approximately the same plane as Saturn's visible ring system (Fig. 7.12). Titan is the only one of Saturn's satellites known to have an atmosphere. Historically, it had been assumed that the composition of Titan's atmosphere was similar to its mother planet, however, observational studies and Voyager 1 have recently shed more light on the specific atmospheric constituents. The Voyager 1 flyby of Titan in November, 1980 discovered that nitrogen comprised 99 percent of Titan's atmosphere. The remaining 1 percent consisted primarily of methane along with trace amounts of other hydrocarbons. Methane had been suspected as the primary constituent of Titan's atmosphere because of the manner in which the atmosphere absorbed ultraviolet energy. This idea had been reinforced by the observation of a hydrogen cloud that extended 5 R_S from Titan's center. Observers suggested that the source of the hydrogen in the cloud resulted from the dissociation of methane (CH_4) into its chemical components (hydrogen and carbon). The lighter hydrogen would slowly escape to space while the carbon aerosols settled toward Titan's surface. Although Titan's atmosphere is primarily nitrogen, the presence of methane supports the ideas concerning the source of the hydrogen in the cloud. It had been assumed that all of the light gases would have escaped from Titan's weak gravitational field. Instead, we now know that Titan's atmosphere is as dense, if not denser, than the earth's atmosphere.

The fact that Titan had a dense, nitrogen atmosphere was an extremely important revelation. The discovery that Titan's atmospheric temperature regime differed from the very low temperatures most scientists had expected to find was perhaps even more significant. The initial studies of Titan's temperature regime revealed that an observed temperature of $93^{o}K$ in the upper atmosphere

Fig. 7.12. Titan and its clouds are seen in this Voyager 1 image taken on November 9, 1980 at a distance of 4.5 million km. (Courtesy Jet Propulsion Laboratory/NASA.)

was significantly lower than would be expected for a slowly rotating moon so distant from the sun. This low temperature implied that Titan's atmosphere must be efficiently trapping the solar energy it received and strongly suggested a greenhouse effect with higher temperatures at lower altitudes.

Early estimates expected the data to show Titan's maximum energy emission in the infrared wavelengths at 20-30 μm. These studies found that much less energy was radiated at these wavelengths than absorbed, and concluded that the atmosphere must be opaque. Today, we know the atmosphere is opaque. It was again assumed that the trapped energy heated the surface via a greenhouse effect. The existence of the greenhouse effect was initially substantiated by temperature measurements which showed that the bulk of Titan's infrared radiation was emitted at 8.4 μm, 11.0 μm, and 12.0 μm at temperatures of 145°K, 135°K, and 125°K, respectively. More recent data from the Voyager 1 spacecraft suggest Titan's temperature is ≈80°K, lower than had been previously anticipated.

Sagan attributed Titan's greenhouse effect to the presence of molecular hydrogen, but noted that there are problems with the quantities involved. Most scientists felt that large quantities of hydrogen were necessary to produce the effect, yet large quantities of hydrogen had not been observed on Titan. Sagan suggested that the presence of hydrogen in Titan's atmosphere was masked by dense clouds. Polarization measurements by other researchers confirmed the existence of a dense cloud cover around Titan even prior to Voyager 1 (Fig. 7.13). Since Titan's atmosphere is opaque, it is possible that hydrogen continues to remain an issue. Sagan suggested that the hydrogen came from outgassing on Titan and suggested that volcanic action was the mechanism whereby methane, ammonia, and water were released into Titan's atmosphere. Once in the atmosphere, they were broken down into molecular hydrogen and organic compounds. These compounds would produce a greenhouse effect and contribute to surface temperatures near 200°K. There are, of course, no known volcanoes on Titan, however, we do know they exist on Io. If tidal heating is responsible for Io's volcanoes, then Titan may very well have volcanoes as a similar set of parameters exist between Saturn and Titan. Titan is a relatively large moon and is subject to the gravitational and magnetic forces of a rapidly rotating Saturn. Titan's geologic composition will probably be the final determinant of volcanism on Titan. Io's volcanic cycle is related to the pressurization and heating of sulfur compounds. We simply do not know what Titan's geology is like and any suggestions at this time are purely speculative. We do know that the recent data suggests that the proposed 200°K surface temperature is probably an extremely high estimate.

There are some vexing problems raised by Titan's characteristics. Light gases such as hydrogen should have long since escapted Titan's atmosphere. A number of suggestions have been offered to explain the

Fig. 7.13. Titan's thick haze layer is shown in this enhanced Voyager 1 image taken on November 12, 1980 at a distance of 435,000 km. Titan is enveloped in a thick haze which merges with a darker "hood" or cloud layer over its north pole. The polar hood is not present at the south pole. (Courtesy Jet Propulsion Laboratory/NASA.)

source of the hydrogen. The dissociation of methane has already been mentioned. Another explanation suggests that the hydrogen is not continuously produced but is recycled. While Titan's hydrogen does escape, its escape velocity is not sufficiently large enough to escape Saturn's gravitational field. It was postulated and confirmed that the hydrogen would accumulate to form a doughtnut-shaped ring, i.e. a torus, around Saturn in the plane of Titan's orbit. Titan would eventually recapture most of the lost hydrogen as it orbited Saturn. Titan and Io are therefore unique in that they may be the only moons in our solar system capable of cycling gas molecules from their atmosphere into outer space and later retrieving them.

Clouds

The nitrogen and methane in Titan's atmosphere are probably accompanied by varying amounts of hydrogen and ammonia. Titan's atmosphere ranges in color from yellow to red. These colors could more easily be explained if Titan had a layer of ammonium hydrosulfide clouds. Ammonium hydrosulfide is yellow in color when bombarded by ultraviolet radiation from the sun. The reddish colors are thought to result from the photodissociation of methane and possibly ammonia and water. The photodissociation of nitrogen gases results in a yellowish-brown haze, similar to urban photochemical smog on earth. These photodissociation processes would be sufficient to explain the range of colors found in Titan's atmosphere.

Interestingly enough, it has been suggested that these same photodissociation processes might create a surface layer of hydrocarbons on Titan. The carbon dissociation from the methane could combine to produce many different organic molecules. Attempts to model this process suggest that polyethylene, a polymer of methane, might fall to Titan's surface. Other scientists suggest that the photodissociation of methane creates amino acids, the building blocks of life. Certainly, the existence of amino acids on Titan should be considered highly speculative at best.

Atmospheric Pressure

Surface atmospheric pressure on Titan has been estimated as at least equal to the earth's atmospheric sea level pressure (≃1,000 mb). The actual value of Titan's surface pressure should become available after the Voyager 2 flyby in August, 1981.

Life on Titan

The question of life on Titan has been raised by many researchers. Surface temperatures on Titan are cold. Titan may also have a surface with pools of nitrogen and methane into which photoinduced organic compounds, precipitated out of the atmosphere, fall. These compounds may provide conditions supportive to the development of life. Titan may therefore be a modern day laboratory from which scientists may gain insight into the evolution of life or its precursors, however, the existence or development of "life" generally requires that it occur in environments where water is available. Water's unique characteristics of low boiling and high freezing point, its thermal capacity, and the ease with which the hydrogen bonds are broken to release a hydrogen ion make it truly the universal solvent. Chemical reactions forming complex molecules easily occur in water. The problem is that water appears to be absent on Titan.

Temperature is another factor affecting the rates at which chemical reactions occur and hence the probability of the formation of life associated compounds. Perhaps, unfamiliar life forms might develop in pools of nitrogen, methane, and ammonia and that conditions comparable to reactions which occur in water on earth might exist on Titan. Certainly, the continued search for life is very interesting and should be pursued. It does appear that Titan is our last hope of finding "life forms" or the precursors of life in our solar system.

Summary

Titan has emerged as one of the more interesting moons in our solar system. Its atmosphere, composition, temperature, recycling of hydrogen with its torus, and environmental conditions which might support the evolution of life forms make it a target for future research.

References

Acuna, M.H. and Ness, N.F., The magnetic field of Saturn, Pioneer 11 observations, Science, 207, 444 ff., 1980.

Cruikshank, D.P. and Morrison, D., Titan and its atmosphere, Sky and Telescope, 44, 83 ff., 1972.

Dyer, J.W., Pioneer Saturn, Science, 207, 400 ff., 1980.

Fillius, N. et al., Trapped radiation belts of Saturn: first look, Science, 207, 425 ff., 1980.

Gehrels, T. et al., Imaging photopolarimeter on Pioneer Saturn, Science, 207, 434 ff., 1980.

Hunten, D.M., The outer planets, Scientific American, 233, 130 ff., 1975.

Ingersoll, A.P. et al., Pioneer Saturn infrared radiometer: preliminary results, Science, 207, 439 ff., 1980.

Judge, D.L. et al., Ultraviolet photometer observations of the Saturnian system, Science, 207, 431 ff., 1980.

Kliore, A.J. et al., Vertical structure of the ionosphere and upper neutral atmosphere of Saturn from the Pioneer radio occultation, Science, 207, 446 ff., 1980.

NASA/JPL, Mission status reports, Voyager Bulletin, 51-60, 1980.

Opp, A.G., Scientific results from the Pioneer Saturn encounter: summary, Science, 207 401 ff., 1980.

Simpson, J.A. et al., Saturnian trapped radiation and its absorption by satellites and rings: the first results from Pioneer 11, Science, 207, 411 ff., 1980.

Smith, E.J. et al., Saturn's magnetic field and magnetosphere, Science, 207, 407 ff., 1980.

Trainor, J.H. et al., Observations of energetic ions and electrons in Saturn's magnetosphere, Science, 207, 421 ff., 1980.

Van Allen, J.A. et al., Saturn's magnetosphere, rings and inner satellites, Science, 207, 415 ff., 1980.

Wolfe, J.H. et al., Preliminary results on the plasma environment of Saturn from the Pioneer 11 plasma analyzer experiment, Science, 207, 403 ff., 1980.

CHAPTER VIII

URANUS, NEPTUNE, AND PLUTO

Introduction

Historically, research on the three outer planets has been neglected due to their inaccessibility at the outer limits of our solar system. The discovery of nine rings lying in or near the equatorial plane of Uranus and the anticipation of the flybys of the Voyager spacecraft has generated renewed interest in Uranus, Neptune, and Pluto. Scientifically, very little is known about the outer planets. The 1986 Voyager flyby of Uranus and later rendezvous with Neptune (1989) may change all this (if the spacecraft survives). Initially, it had been assumed that Uranus and Neptune were probably scaled-down versions of Jupiter and Saturn. Recent discoveries have shown that they are quite different from Jupiter and Saturn.

URANUS

Descriptive Statistics:

Radius	25,900 km
Distance from the Sun	19.2 A.U.
Period of Orbit	84 years
Period of Rotation	22 ± 2 hours (retrograde)
Axial Tilt	98°
Mass	14.5 times earth
Density	1.2 g cm^{-3}

Atmospheric Composition

Spectra reveal that Uranus' atmosphere is primarily composed of methane and hydrogen, although the ratio between the two is unknown. Uranus appears to be noticably lacking in helium although atmospheric models suggest that it may be present. Similarly, ammonia is also a suspected atmospheric constituent that has not been detected as yet. It is possible that ammonia is present but locked up as an ammonium hydrosulfide cloud layer deeper in the atmosphere. Suggestions for a methane cloud layer have also been made, however, at the present time, it is impossible to prove or disprove either the ammonium hydrosulfide or the methane cloud hypothesis.

Temperature

Atmospheric temperature measurements are difficult to obtain when the planet is 19.2 A.U. from the sun. Infrared measurements from the 10^2 mb pressure level show an effective temperature of 58°K. Temperature measurements taken from the 10^{-3} mb pressure level during Uranus' occultation of a star show a temperature of 100°K. A temperature/pressure diagram for two locations, based on occultation data, are presented in Fig. 8.1. The mean upper atmospheric

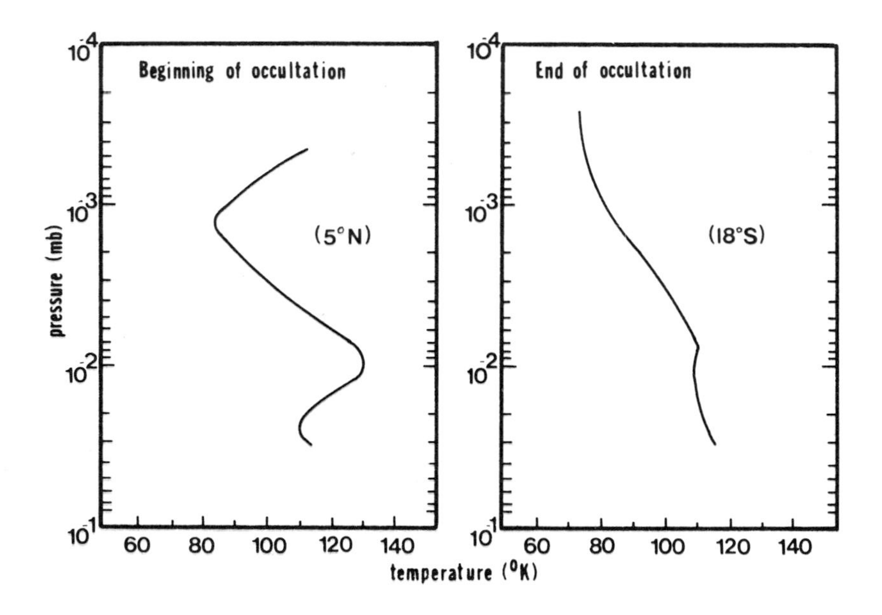

Fig. 8.1. Temperature/pressure diagram for Uranus' atmosphere. (After J.L. Elliot and E. Dunham, Nature, v 279, 307 ff., 1979. Copyright 1979 by the Macmillin Journals Ltd.)

temperature between the 10^{-2}-10^{-3} mb pressure level is 100°K. The occultation temperature of the upper atmosphere is larger than the infrared temperatures from a lower atmospheric level. There is general agreement that the 40°K difference indicates the existence of a temperature inversion. The inversion may be associated with the process of methane dissociation. The photodissociation of methane would absorb solar energy at some altitude above the planet. The increase in temperature would explain the inversion layer, however, this is only speculation for the present. The overall temperature structure of the Uranian atmosphere is subadiabatic and, as a result, drastically limits the potential for convective mixing and turbulence.

Speculation on an internal heat source for Uranus is based on the infrared data which suggested Uranus radiated 1.2 times more energy than it received from the sun. The theoretical black body radiation for a planet of this size and distance from the sun negates the internal heat source hypothesis and strongly suggests that Uranus may lack one altogether. It is impossible to determine which idea may be correct without additional data, however, the effective temperature of 58°K is approximately "normal" for Uranus' size and location in the solar system.

Magnetosphere

Uranus' atmosphere contains less hydrogen than Saturn's atmosphere. Helium is not present at all. This combination of characteristics suggests that the planet may contain more of the denser elements, leading to speculation that Uranus might have a magnetic field and a magnetosphere. Direct evidence for a magnetic field and magnetosphere is lacking and will probably remain so for some time to come.

The number of unanswered questions about Uranus continue to accumulate. What is the nature of the atmospheric heating and the inversion level? Is it produced by methane absorption of solar energy or is it the result of some unique process? Does the planet have its own internal heat source like its neighbors, Jupiter, Saturn, and Neptune? Is a magnetosphere present or does the upper Uranian atmosphere compress under the influence of the solar wind in a manner similar to Venus' ionosphere? These and many more questions will probably remain unanswered until the Voyager probe reaches Uranus in 1986.

NEPTUNE

Descriptive Statistics:

Radius	24,750 km
Distance from the Sun	30.1 A.U.
Period of Orbit	164 years

Period of Rotation	21 ± 2 hours
Mass	17 times earth
Density	2.2 g cm^{-3}

Atmospheric Composition

Neptune's atmosphere seems to be composed of methane and hydrogen, the same constituents that have been identified in Uranus' atmosphere. Helium and ammonia have not been detected but are thought to be present in Neptune's atmosphere. One gas, ethane, was detected on Neptune but not on Uranus.

Temperature

The effective infrared temperature at the 10^2 mb pressure level is 60^oK. Occultation temperature data place the temperature at 104^oK at the 10^{-3} mb pressure level. These measurements strongly suggest an internal heat source for Neptune. Its black body temperature at its distance from the sun shoud be $\approx 45^oK$. Measurements show that Neptune radiates 2.8 times as much energy as it receives from the sun, however, scientists are at a loss to explain the mechanism or process responsible for the surplus heat energy.

Albedo

From early 1975 to early 1976, Neptune brightened, i.e. appeared to reflect more efficiently. By reflecting long wave energy in the 1-4 μm range that normally would have been absorbed by methane in the lower atmosphere, Neptune's brightness increased fourfold, particularly in the infrared range. These variations must represent significant changes in Neptune's cloud layers, confirming both the existence of the cloud layers and their temporal variability. If correct, brightening would represent the first direct observations of meteorological phenomena on Neptune.

The composition of Neptune's clouds is pure conjecture. The similarities with Uranus suggest that they may be composed of methane or ammonium hydrosulfide. More research may provide the answers but they will not come easily because of the vast expanse between the earth and Neptune. Hopefully, the Voyager probe will provide many answers to our questions in 1989.

PLUTO

Descriptive Statistics:

Radius	1,350 km
Distance from the Sun	39.4 A.U.
Period of Orbit	248.4 years
Period of Rotation	6 days, 9 hours, 18 minutes
Density	1.5 g cm^{-3}
Mass	1/380 of earth

Atmospheric Composition and Albedo

Pluto is probably composed of frozen volatiles, i.e. water ice, methane, carbon dioxide, ammonia, etc. Whether or not Pluto has an atmosphere is unknown at the present time. There are no data to suggest that an atmosphere does envelop the planet. Pluto's surface albedo is $\simeq$0.6 (60%). Albedo variations have been observed on Pluto. Normally, albedo variations would be considered evidence for an atmosphere, however, in Pluto's case, these variations are attributed to surface variations in the distribution of the volatile frost.

Temperature

The temperature on Pluto is probably $>40^{o}$-50°K. This value is based on the temperatures at which methane condenses into a solid and would support the notion that Pluto is coated with a layer of methane frost.

Summary

So little is known about Pluto that even the discovery of its correct radius and the existence of a satellite were significant additions to our knowledge of the planet. Typically though, this information raised an additional, unanswered question. Earlier estimates had determined that Pluto's radius was twice its actual value. This was sufficient to explain the perturbations observed in Uranus' and Neptune's orbits. The new data suggest that Pluto is much too small to exert any measurable effect on Uranus and Neptune. The data suggest another mass, farther away from the sun than Pluto. Is there a tenth planet or is the mass of the cometary cloud, thought to exist beyond Pluto, sufficient to exert these known effects? The answer to this and the many other questions will tax the imaginations, curiosity, and resources of future planetary scientists.

References

Cruikshank, D.P. et al., Pluto: evidence for methane frost, Science, 194, 835 ff., 1976.

Elliot, J.L. and Dunham, E., Temperature structure of the uranian upper atmosphere, Nature, 279, 307 ff., 1979.

Hubbard, W.B. and MacFarlane, J.J., Structure and evolution of Uranus and Neptune, J. Geophys. Res., 85, 225 ff., 1980.

Hunt, G.E., Weather on Neptune, Nature, 269, 10 ff., 1977.

Hunt, G.E., Atmospheres of Uranus and Neptune, Nature, 272, 403 ff., 1978.

Hunten, D.M., The outer planets, Scientific American, 233, 130 ff., 1975.

Hunten, D.M., New surprises from Uranus, Nature, 276, 16 ff., 1978.

Kerr, R., Rings around the solar system, Science, 306, 38 ff., 1979.

Metz, W.D., Coldest planet: methane ice found on Pluto, Science, 192, 362 ff., 1976.

Millis, R.L. et al., Detection of rings around Uranus, Nature, 267, 330 ff., 1977.

Prinn, R. and Lewis, J., Uranus' atmosphere: structure and composition, The Astrophysical Journal, 179, 333 ff., 1973.

Sinton, W.M., Near infrared view of the Uranus system, Sky & Telescope, 44, 304 ff., 1972.

INDEX